PART I
STUDENT STUDY GUIDE

CALCULUS 7TH
AND ANALYTIC GEOMETRY

THOMAS/FINNEY

MAURICE D. WEIR

NAVAL POSTGRADUATE SCHOOL

ADDISON-WESLEY PUBLISHING COMPANY, INC.
Reading, Massachusetts · Menlo Park, California · New York
Don Mills, Ontario · Wokingham, England · Amsterdam · Bonn
Sydney · Singapore · Tokyo · Madrid · Bogotá · Santiago · San Juan

PREFACE TO THE STUDENT

This study guide has been designed especially for you, the student. It conforms with the seventh edition of CALCULUS AND ANALYTIC GEOMETRY, by George B. Thomas and Ross L. Finney. It is intended as a self-study workbook to assist you in mastering the basic ideas in calculus. Although this manual was written to conform to the Thomas/Finney CALCULUS AND ANALYTIC GEOMETRY, it can be used to accompany any standard calculus textbook and course.

Organization And Learning Objectives

The study manual is organized section by section to correspond with the Thomas/Finney text. For each section we specify its main ideas by stating appropriate learning OBJECTIVES. Each objective states a particular task for you to perform in order to master that objective. Usually the task requires you to solve a certain type of problem related to the discussion in the text; sometimes the task requires you to demonstrate proficiency with certain key terms or concepts. In every case the objective is highly specific and states exactly what you must do.

One or more examples follows each objective and illustrates its requirements. Each example is written in a semi-programmed format; that is, the example is only partially worked out, so you must supply some of the intermediate results yourself. Correct answers to each intermediate result are supplied at the bottom of the page. Thus, each example is broken down into a sequence of steps to guide you through the procedures and techniques associated with its solution. Each example has been carefully selected not to repeat examples or problems in the Thomas/Finney text; thus you retain the full array of the text problems for practice and further learning.

Self-tests

At the end of each chapter there is a SELF-TEST. Each test is followed by complete solutions to all the test problems. The test problems cover the objectives and are similar in scope and difficulty to the examples in this manual and the examples and problems in the Thomas/Finney text. The test should be useful in preparing for class examinations.

How To Use This Study Guide

We recommend that this manual be used in the following way:

1. **Read the textbook**: Carefully read the section of Thomas/Finney assigned you by your calculus instructor.

2. **Study the learning objectives and examples**: Read each objective and work through the associated example(s) in the corresponding section of this manual. You should conceal the answers to the examples at the bottom of the page. Work with pencil and scratch paper as you are guided through each solution, writing in the intermediate results in the blanks provided.

3. **Check your answers**: After all the blanks for a given problem are filled in, compare your answers with the correct answers at the bottom of the page. If you have difficulty or do not fully understand the answers given, review the material in the textbook or consult with your instructor.

4. **Do the chapter self-test**: After you complete a chapter in Thomas/Finney, review the objectives for that chapter in this manual. Then take the chapter self-test and compare your solutions with those provided. Problems in the self-test sometimes bring together several ideas from the chapter.

Guidance From Your Instructor

We caution you that the learning objectives given in this manual by no means exhaust all the possible objectives that could be written for a careful study of calculus; we have tried to identify the main ones. However, your instructor may have additional requirements. For instance, he or she may want you to be able to prove certain theorems or derive results in the text. We have not stated objectives of this sort. Also, your instructor may consider some objectives far more important than others and not require that you master some objectives at all. So it is imperative that you find out specifically what your instructor considers essential, and study accordingly. This manual should be helpful to you both in identifying the tasks and successfully mastering them. The problems assigned by your instructor should help you discover those concepts and applications of calculus that your instructor wishes to stress.

Maurice D. Weir

TABLE OF CONTENTS

Chapter 1
THE RATE OF CHANGE OF A FUNCTION   1

Chapter 2
DERIVATIVES   27

Chapter 3
APPLICATIONS OF DERIVATIVES   45

Chapter 4
INTEGRATION   69

Chapter 5
APPLICATIONS OF DEFINITE INTEGRALS   87

Chapter 6
TRANSCENDENTAL FUNCTIONS   105

Chapter 7
METHODS OF INTEGRATION   125

Chapter 8
CONIC SECTIONS AND OTHER PLANE CURVES   145

Chapter 9
HYPERBOLIC FUNCTIONS   165

Chapter 10
POLAR COORDINATES   175

Chapter 11
INFINITE SEQUENCES AND INFINITE SERIES   187

Chapter 12
POWER SERIES   207

CHAPTER 1 THE RATE OF CHANGE OF A FUNCTION

PROLOGUE.

OBJECTIVE : Broadly define the mathematics of calculus and specify
its two main classes of problems.

1. Calculus is the mathematics of _motion_ and _change_ .

2. One class of problems in calculus involves finding the
rate at which a variable quantity is changing. This
branch of calculus is called the _differential_ calculus.

3. A second class of problems in calculus involves determining a
function when its rate of change is known. This branch is
called the _integral_ calculus.

Both branches are important to modern science and engineering.

1-1 COORDINATES FOR THE PLANE.

OBJECTIVE : Draw a rectangular coordinate system and plot or locate
points within it.

4. Finish labeling the coordinate
system at the right and plot
the point

$$P = P(-3,2).$$

$(2,2)$ '$(2,-2)$
$(-3,-2)$

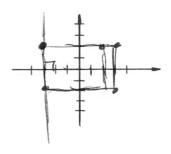

5. Use the same diagram to locate the point Q such that PQ is
perpendicular to the x-axis and bisected by it. The
coordinates of Q are $(-3,-2)$.

---

1. change, motion

2. rate, differential

3. function, integral

4. P(-3,2)

5. Q(-3,-2)

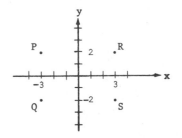

6.   Use the same diagram to locate the point R   such that PR   is
     perpendicular to the y-axis and bisected by it.   The
     coordinates of R   are ___(3, 2)___ .

7.   Use the same diagram to locate the point S   such that PS   is
     bisected by the origin.   The coordinates of S   are ___(3, -2)___ .

1-2   THE SLOPE OF A LINE.

OBJECTIVE A :   Given the coordinates of points P   and   Q   in the
                plane, find the increments Δx and Δy from P
                to Q   and the distance between the points.

8.   If a particle starts at   P(2,-1)   and goes to   Q(-7,-3),   then
     its x-coordinate changes by

$$\Delta x = -7 - \underline{2} = \underline{-9} .$$

9.   Its y-coordinate changes by

$$\Delta y = \underline{-3} - (-1) = \underline{-2} .$$

10. The distance between P   and Q   is

$$d = \sqrt{\underline{-9^2} + \underline{-2^2}} = \underline{\sqrt{85}} .$$
$$81 + 4$$

OBJECTIVE B :   Given the increments from the point P   to the point Q
                and the coordinates of one of these points, determine
                the coordinates of the other point.

11. The coordinates of a particle change by $\Delta x = -3$ and $\Delta y = 5$
     in moving from   P(1,-4)   to   Q(x,y).   The x-coordinate of Q
     is given by
$$1-3$$
$$x = 1 + \underline{-3} = \underline{-2} .$$

12. The y-coordinate of Q   is given by

$$y = \underline{-4} + 5 = \underline{1} .$$

---

6. R(3,2)   7. S(3,-2)   8. 2, -9   9. -3, -2

10. $(-9)^2$, $(-2)^2$, $\sqrt{85}$ 11. -3, -2   12. -4, 1

OBJECTIVE C : Define the slope of a straight line and calculate the slope (if any) of the line determined by two given points.

13. The slope of the line through the points $P_1(x_1, y_1)$ and $P_2(x_2, y_2)$ is given by

$$m = \frac{rise}{run} = \frac{y_2 - y_1}{x_2 - x_1} = \frac{y_1 - y_2}{x_1 - x_2}, \text{ provided that } x_1 \neq x_2 .$$

14. If $x_1 = x_2$, then the line through the points $P_1(x_1, y_1)$ and $P_2(x_2, y_2)$ is a *vertical* line. For vertical lines, the *slope* is not defined.

15. The slope of the line through the points $A(-\frac{1}{2}, 1)$, $B(0, -2)$ is
$m = \underline{\ -6\ }$.

$$\frac{-2 - 1}{0 + \frac{1}{2}} = \frac{-3}{\frac{1}{2}} \quad -6$$

OBJECTIVE D : Find the slope (if any) of a line perpendicular to a line determined by two given points.

16. The slope of the line perpendicular to AB in Problem 15 is
$m = \underline{\ \frac{1}{6}\ }$.

OBJECTIVE E : Use slopes to determine whether three or more points are collinear (lie on a common straight line).

17. Consider the three points $A(-3, 7)$, $B(1, -1)$, $C(2, -3)$. The slope of the line through A and B is

$$m_1 = \underline{\ -2\ } .$$

$$\frac{-1 - 7 = -8}{1 + 3 = 4}$$

The slope of the line through B and C is

$$m_2 = \underline{\ -2\ } .$$

$$\frac{-3 + 1 = -2}{2 - 1 = +1}$$

Because m_1 and m_2 are $\underline{\quad = \quad}$, the three points A, B, C are collinear.

1-3 EQUATIONS FOR LINES.

OBJECTIVE A : Write an equation of any vertical line given a point on the line.

18. An equation of the vertical line passing through the point $P(4, -7)$ is $\underline{\ x = 4\ }$.

---

13. $\frac{y_2 - y_1}{x_2 - x_1}, \frac{y_1 - y_2}{x_1 - x_2}$ 14. vertical, slope   15. -6   16. $\frac{1}{6}$

17. $m_1 = -2$, $m_2 = -2$, equal   18. $x = 4$

OBJECTIVE B : Write an equation of any line with given slope and passing through a given point.

19. Using the point-slope equation of the line, we have $y - y_1 = m(x - x_1)$. Thus an equation of the line with slope $m = -2$ through the point $(1,3)$ is given by _____ .

20. The line perpendicular to the line in (19) has slope $m = -\frac{1}{-2} = \frac{1}{2}$. Thus an equation of the perpendicular through $(1,3)$ is _____ .

OBJECTIVE C : Write an equation of any line given two points on the line.

21. Let $P_1(-3,0)$ and $P_2(2,-1)$ be two points on the line L. The slope of L is $m =$ _____ . Thus, an equation of L using P_1 is _____ , using P_2 an equation is _____ . In either case, solving for y we obtain the equation $y =$ _____ .

22. Let $P_1(1,-3)$ and $P_2(1,5)$ be two points on the line L. Since the x-coordinates of the points are the same, we conclude that L is a _____ line and hence has no _____ . An equation for L is _____ .

OBJECTIVE D : Recognize an equation as representing a line and determine the slope (if any), the x-intercept (if any), and the y-intercept (if any).

23. The equation $3x - 2y = 6$ represents a straight line because it contains only _____ powers of x and y. When $x = 0$, $y =$ _____ which gives the value where the line crosses the y-axis. This is called the _____ . When $y = 0$, $x =$ _____ giving the value where the line crosses the _____ . This is called the x-intercept.

24. The equation $y = 3$ represents a straight line that is parallel to the _____ . It is called a _____ line and has slope $m =$ _____ .

25. The equation $xy = 1$ does not represent a straight line because it is not a _____ equation when the variables x and y are multiplied together.

---

19. $y - 3 = -2(x - 1)$

20. $y - 3 = \frac{1}{2}(x - 1)$

21. $-\frac{1}{5}$, $y - 0 = -\frac{1}{5}(x + 3)$, $y + 1 = -\frac{1}{5}(x - 2)$, $y = -\frac{x + 3}{5}$

22. vertical, slope, $x = 1$

23. first, -3, y-intercept, 2, x-axis

24. x-axis, horizontal, 0

25. linear

OBJECTIVE E : Graph any equation representing a line.

26.   Graph the line $y = -3x + 1$.

27.   Graph the line $\frac{x}{2} - \frac{y}{3} = \frac{1}{2}$.

OBJECTIVE F : Find an equation of the line passing through a given point and parallel or perpendicular to a given line.

28.   The line containing the point $(-1,2)$ that is parallel to the line $3x - y - 1 = 0$ has slope $m = $ _____ . Since the line contains the point $(-1,2)$, its equation in point-slope form is _____ .

29.   The line containing the point $(4,1)$ that is perpendicular to the line $2y - 3x = 5$ has slope $m = $ _____ . Since the line contains the point $(4,1)$, its equation in point-slope form is _____ .

OBJECTIVE G : Find the distance between two given points in the plane.

30.   The distance between the points $P(-1,3)$ and $Q(2,-5)$ is

$d = \sqrt{(-1 - \text{\_\_\_\_})^2 + (3 - \text{\_\_\_\_})^2} = \text{\_\_\_\_\_}$.

1-4   FUNCTIONS AND GRAPHS.

In this section of the textbook, there are several terms associated with the concept of a function with which you will need to become familiar. The following items are designed to assist you in learning the precise mathematical meanings of these various terms.

31.   A variable is a symbol such as x , y , t , etc. that may take on any value over a prescribed set. The set of values that a variable may take on is called the _____ of the variable.

26.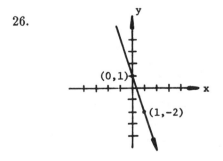

27.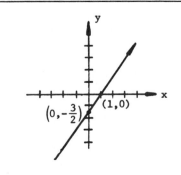

28. 3 , $y - 2 = 3(x + 1)$ 29. $-\frac{2}{3}$, $y - 1 = -\frac{2}{3}(x - 4)$ 30. 2 , -5 , $\sqrt{73}$ 31. domain

32. In applications variables often have domains that are intervals. Symbolically, the open interval a < x < b is given by _____ . The closed interval a ≤ x ≤ b is designated by _____ , and the half-open intervals a < x ≤ b and a ≤ x < b are designated by _____ and _____ , respectively.

33. Calculus is concerned with how variables are related. If to each value of the variable x there corresponds a unique value of the variable y, then y is said to be a _____ of x. The key word in this definition of function is _____ : we do not want to input a single value for the variable x with two or more possible outcomes for y. Every function is determined by two things: (1) the _____ of the first variable x and (2) the _____ or condition describing how y is obtained from x so that the ordered pairs (x,y) belong to the function. The variable x is called the _____ variable or _____ of the function; the second variable y is called the _____ variable. It is also said that the function _____ the x-variable to its image y-value. The set of values taken on by the dependent variable y is called the _____ of the function.

OBJECTIVE A : Given an equation for a function y = f(x) calculate the value of f at a specified point, find the domain and range of f, and graph f by making a table of pairs.

34. Consider the function y = -3x + 2. The domain of the function is the interval _____ . Solving the equation for x, gives x = _____ so that the variable y may take on any value whatsoever. Thus, the range of the function is the interval _____ . Sketch the graph.

35. Consider the function $y = \frac{x^2 - 1}{x + 1}$. The function is defined for all values of x except _____; hence the domain consists of the union of the intervals _____ and _____ . When x ≠ -1, $y = \frac{x^2 - 1}{x + 1} = \frac{(x - 1)(x + 1)}{x + 1} =$ _____ . Therefore, the range of the function is all real numbers except for y = _____ (because x ≠ -1), so that the range is the union of the two intervals _____ and _____ .

---

32. (a,b), [a,b], (a,b], [a,b)

33. function, unique, domain, rule, independent, argument, dependent, maps, range

34. (-∞,∞) , $x = \frac{2 - y}{3}$, (-∞,∞)

35. x = -1
(-∞,-1) and (-1,∞)
x - 1,   -2,
(-∞,-2) and (-2,∞)

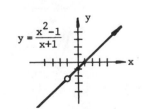

36. The domain of the function $y = -\sqrt{1 - x}$ is the interval
_____, since $\sqrt{1 - x}$ is defined whenever $1 - x \geq 0$.
Squaring both sides and solving the resultant equation for x,
we obtain $x =$ _____. We see from this last equation
that y can take on any value. However, since y is the
negative square root, the range is the interval _____.

37. If $g(x) = \dfrac{1}{\sqrt{x - 2}}$, the domain of g is _____. The value
$g(3)$ is _____; $g(11)$ is _____; $g(a)$ is _____;
$g(b + 2)$ is _____.

OBJECTIVE B : Given two functions f and g, write an expression
for their composite $f(g(x))$.

38. If $f(x) = 5x + 2$ and $g(x) = x^2$, then a formula for $f(g(x))$
is obtained as follows:

$$f(g(x)) = f(x^2) = \underline{\hspace{2cm}}.$$

The domain of $y = f(g(x))$ is all values of x in the domain
of g such that $f(g(x))$ is defined. This is the interval
_____.

39. If $f(x) = \sqrt{x - 1}$ and $g(x) = x + 1$, then

$$f(g(x)) = \underline{\hspace{2cm}}.$$

The domain of the composite is all values of x in the domain
of g such that $f(g(x))$ is defined. This is the interval
_____.

40. Let $f(x) = x^2$ and $g(x) = \sqrt{x - 1}$. Then $f(g(x)) =$ _____.
The domain of g is the set of all real numbers x satisfying
_____. Thus, the domain of the composite $y = f(g(x))$
is the interval _____.

---

36. $(-\infty, 1]$,

$x = 1 - y^2$

$(-\infty, 0]$

$y = -\sqrt{1-x}$

37. $(2, \infty)$, 1, $\frac{1}{3}$, $\dfrac{1}{\sqrt{a - 2}}$, $\dfrac{1}{\sqrt{b}}$

38. $5x^2 + 2$, $(-\infty, \infty)$

39. $\sqrt{x}$, $[0, \infty)$

40. $x - 1$, $x \geq 1$, $[1, \infty)$

OBJECTIVE C : Find two functions f and g that will produce a given composite function h such that $h(x) = f(g(x))$.

41. Consider $h(x) = \sin(x^2 - 1)$. If we let $f(x) = \sin x$ and $u = g(x) =$ _____ , then
$h(x) =$ _____ $= f(x^2 - 1) =$ _____ .

42. If $h(x) = \sqrt{x^5 + 2x^3 - 1}$, then for $f(x) = \sqrt{x}$ and $u = g(x) =$ _____ , it is true that $h(x) = f(g(x))$.

1-5   ABSOLUTE VALUES.

OBJECTIVE A : Define <u>absolute value</u> and describe the domain of an absolute value inequality without using absolute value symbols.

43. The absolute value function assigns to the number x the number _____ . Thus, if $x \geq 0$ so that x is nonnegative, $|x|$ is the number _____ ; but if $x < 0$ is negative, then $|x|$ is the number _____ . For instance, $|4| =$ _____ and $|-4| = - ($ ___ $) =$ _____ . Thus, the absolute value function is never negative. Its domain is the interval _____ and its range is the interval _____ .

44. The number _____ = _____ measures the distance between x and a. If r is a positive real number, then $|x-a| < r$ is equivalent to the inequality _____ . Thus, x must lie within the interval _____ .

45. If $|x+2| \leq 7$, then $|x-(-2)| \leq$ _____ . This is equivalent to the inequality _____ . Therefore x must lie within the closed interval _____ .

46. If $|x| > r > 0$, then x must lie within the union of the intervals _____ and _____ . For instance, $|x| > 7$ implies $x <$ _____ or $x >$ _____ .

47. The relationship between absolute value, addition, and multiplication is given by the two equations

$|a+b| \leq$ _____

$|ab| =$ _____ .

---

41. $x^2 - 1$, $f(g(x))$, $\sin(x^2 - 1)$ 42. $x^5 + 2x^3 - 1$ 43. $\sqrt{x^2}$, x, -x, 4, -4, 4, $(-\infty,\infty)$, $[0,\infty)$

44. $|x - a| = |a - x|$, $-r < x - a < r$ or $a - r < x < a + r$, $(a - r, a + r)$

45. 7, $-7 \leq x+2 \leq 7$, $[-9,5]$ 46. $(-\infty,-r)$ and (r,∞), -7, 7

47. $|a| + |b|$, $|a||b|$

OBJECTIVE B : Given a function involving absolute value or the greatest integer in x, describe the domain of the function and graph the function.

48. Consider the function y = |x - 1| + 2. The domain of this function is the interval _____ . If the x values satisfy x ≥ 1, then y = _____; on the other hand, if x < 1, then y = _____ . A table of some values for this function is given by (complete the table):

x	-2	-1	0	1	2	3
y	5					

Sketch the graph of the function using the table. From the graph the range of the function is evidently the interval _____ .

49. Consider the function y = [x - 1] + 2, where [x - 1] denotes the greatest integer in _____ . The domain of this function is the interval _____ . A table of some of the values for this function is given by (complete the table):

x	-2.0	-1.5	-1.0	-.5	0	.5	1.0	1.5	2.0	2.5
y	-1.0									

Sketch a graph of the function using the table. The range of this function is not an interval, but the set of numbers _____ .

---

48. (-∞,∞),

(x - 1) + 2 = x + 1,

-(x - 1) + 2 = 3 - x,

x	-2	-1	0	1	2	3
y	5	4	3	2	3	4

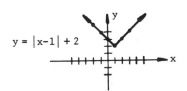

y = |x-1| + 2

range: [2,∞)

49. x - 1,

(-∞,∞),

x	-2.0	-1.5	-1.0	-.5	0
y	-1.0	-1.0	0	0	1.0

x	.5	1.0	1.5	2.0	2.5
y	1.0	2.0	2.0	3.0	3.0

y = [x-1] + 2

range: {...,-2,-1,0,1,2,3,...}

1-6   TANGENT LINES AND THE SLOPES OF QUADRATIC AND CUBIC CURVES.

OBJECTIVE A : Use the method of finding the limit of m_{sec} as Δm approaches zero to find a formula that gives the slope at any point $P(x,y)$ on a given quadratic or cubic curve.

50.   Consider the curve $y = 6 - 4x - x^2$. If $P(x_1,y_1)$ is a point on the curve and $Q(x_2,y_2)$ is another point on the curve with

$$\Delta x = x_2 - x_1 \quad \text{and} \quad \Delta y = y_2 - y_1 \; ,$$

then $x_2 =$ _____ and $y_2 =$ _____ . Since the point $Q(x_2,y_2)$ is on the curve,

$$y_2 = 6 - 4x_2 - x_2^2 = \underline{\hspace{6cm}}$$
$$= \underline{\hspace{5cm}} \; .$$

Since P is on the curve, its coordinates also satisfy the equation:

$$y_1 = \underline{\hspace{4cm}} \; .$$

Thus, $\Delta y = y_2 - y_1 = \underline{\hspace{4cm}} \; .$
Division of both sides by Δx gives,

$$m_{sec} = \frac{\Delta y}{\Delta x} = \underline{\hspace{4cm}} \; .$$

The slope of the tangent to the curve at the point (x_1,y_1) is the _____ of m_{sec} as Δx approaches _____ . Hence, this limit equals _____ . Since (x_1,y_1) is an arbitrary point on the curve, deleting the subscript 1 gives $m =$ _____ , the slope at any point $P(x,y)$.

OBJECTIVE B : Find an equation for the tangent to a given quadratic or cubic curve at a specified point on the curve.

51.   For the curve $y = 6 - 4x - x^2$ in the preceding Problem 50, let's find an equation of the tangent line at the point $(-1,1)$. The slope of the tangent line was found previously to be $m = -4 - 2x$. Thus, at $x = -1$, $m =$ _____ . The point-slope equation for the tangent line is therefore:

$$y - \underline{\hspace{2cm}} = -2(x - \underline{\hspace{2cm}}) \; ,$$

or

$$y = \underline{\hspace{4cm}} \; .$$

---

50. $x_1 + \Delta x$, $y_1 + \Delta y$, $6 - 4(x_1 + \Delta x) - (x_1 + \Delta x)^2$, $6 - 4x_1 - 4\Delta x - x_1^2 - 2x_1\Delta x - (\Delta x)^2$,

$6 - 4x_1 - x_1^2$, $-4\Delta x - 2x_1\Delta x - (\Delta x)^2$, $-4 - 2x_1 - \Delta x$, limit, zero, $-4 - 2x_1$, $m = -4 - 2x$

51. -2, 1, -1, $-2x - 1$

1-7   THE SLOPE OF THE CURVE $y = f(x)$,   DERIVATIVES.

OBJECTIVE A : For a given function f, find the derivative $f'(x)$ by applying the definition.

52.   The definition of the derivative of f at the point x_1 is

$$f'(x_1) = \lim_{\Delta x \to 0} \frac{f(x_1 + \Delta x) - f(x_1)}{\Delta x}$$

whenever this limit exists. The set of all pairs of numbers $(x, f'(x))$ is called the *derivative* or *derived* function. The domain of f' is a subset of the domain of f, consisting of all numbers in the domain of f at which the *limit* exists.

53.   For the function $f(x) = (x + 1)^2$,

STEP 1.   Form $f(x + \Delta x) = (x + \Delta x + 1)^2$ $x^2 + x\Delta x + x + x\Delta x + \Delta x^2 + \Delta x + x + \Delta x + 1$

$$= x^2 + 2x\Delta x + 2x + \Delta x^2 + 2\Delta x + 1$$

and $f(x) = x^2 + 2x + 1$.

STEP 2.   Subtract $f(x)$ from $f(x + \Delta x)$:

$$f(x + \Delta x) - f(x) = 2x\Delta x + \Delta x^2 + 2\Delta x$$.

STEP 3.   Divide by Δx:

$$\frac{f(x + \Delta x) - f(x)}{\Delta x} = 2x + \Delta x + 2$$.

STEP 4.   Take the limit as $\Delta x \to 0$:

$$f'(x) = \lim_{x \to 0} \frac{f(x + \Delta x) - f(x)}{\Delta x} = 2x + 2$$.

54.   For the function $f(x) = \dfrac{1}{\sqrt{x - 1}}$,

STEP 1.   Form $f(x + \Delta x) = \dfrac{1}{\sqrt{x + \Delta x - 1}}$ and $f(x) = \dfrac{1}{\sqrt{x - 1}}$

STEP 2.   Subtracting $f(x)$ from $f(x + \Delta x)$, and

STEP 3.   Dividing by Δx gives,

$$\frac{f(x + \Delta x) - f(x)}{\Delta x} = \frac{\sqrt{x - 1} - \sqrt{x + \Delta x - 1}}{\Delta x \sqrt{x - 1} \sqrt{x + \Delta x - 1}}$$

---

52. $\lim_{\Delta x \to 0} \dfrac{f(x_1 + \Delta x) - f(x_1)}{\Delta x}$, derived, derivative, limit

53. $x^2 + 2x\Delta x + (\Delta x)^2 + 2x + 2\Delta x + 1$, $x^2 + 2x + 1$, $2x\Delta x + (\Delta x)^2 + 2\Delta x$,

   $2x + \Delta x + 2$, $2x + 2$

STEP 4.   To calculate the limit as $\Delta x \to 0$ we observe that both the numerator and the denominator in the previous expression tend to zero as $\Delta x \to 0$. In an attempt to avoid division by zero, we rationalize the numerator, obtaining:

$$\frac{f(x + \Delta x) - f(x)}{\Delta x} = \frac{\overline{}}{\Delta x \sqrt{x - 1} \sqrt{x + \Delta x - 1}} \cdot \frac{\left(\right)}{\sqrt{x - 1} + \sqrt{x + \Delta x - 1}}$$

$$= \frac{\overline{}}{(\Delta x \sqrt{x - 1} \sqrt{x + \Delta x - 1})(\sqrt{x - 1} + \sqrt{x + \Delta x - 1})}$$

$$= \frac{\overline{}}{(\sqrt{x - 1} \sqrt{x + \Delta x - 1})(\sqrt{x - 1} + \sqrt{x + \Delta x - 1})}$$

Thus, as $\Delta x \to 0$,

$$f'(x) = \lim_{\Delta x \to 0} \frac{f(x + \Delta x) - f(x)}{\Delta x} = \underline{}$$

$$= -\tfrac{1}{2}(x - 1)^{\underline{}}.$$

OBJECTIVE B :   Write an equation of the tangent line to the curve $y = f(x)$ at a specified value $x = a$.

55.   To find an equation of the tangent line to the curve $f(x) = \dfrac{1}{\sqrt{x - 1}}$ when $x = 2$, we first calculate the slope m.

By definition, $m = \underline{}$. From our calculation in the previous Problem 54, that slope has the value $\underline{}$. The point on the curve corresponding to $x = 2$ has coordinates $\underline{}$. Therefore, the point-slope form gives an equation of the tangent line as $\underline{}$.

1-8   VELOCITY AND OTHER RATES OF CHANGE.

OBJECTIVE :   Given a functional relationship $y = f(x)$ between two variables x and y, use the four step differentiation process to calculate the average rate of change and the instantaneous rate of change of y with respect to x.

56.   Every derivative may be interpreted as the instantaneous rate of change of one variable per unit change in the other. If $y = f(x)$, then

$$\frac{\Delta y}{\Delta x} = \underline{}$$

---

54. $\dfrac{1}{\sqrt{x + \Delta x - 1}}$, $\dfrac{1}{\sqrt{x - 1}}$, $\sqrt{x - 1} - \sqrt{x + \Delta x - 1}$, $\sqrt{x - 1} - \sqrt{x + \Delta x - 1}$,

$(\sqrt{x - 1} + \sqrt{x + \Delta x - 1})$, $(x - 1) - (x + \Delta x - 1)$, -1, $\dfrac{-1}{\sqrt{x - 1} \sqrt{x - 1} (2\sqrt{x - 1})}$, $-\tfrac{3}{2}$

55. $f'(2)$, $-\tfrac{1}{2}$, $(2,1)$, $y - 1 = -\tfrac{1}{2}(x - 2)$

is interpreted as the _____ rate of change of y by a change of one unit in _____ . Passage to the limit as $\Delta x \rightarrow 0$ gives

$$\lim_{\Delta x \to 0} \frac{\Delta y}{\Delta x} = \underline{\qquad}$$

as the _____ rate of change of _____ with respect to _____ .

57.   The derviative $f'(x)$ multiplied by Δx gives the change that would occur in _____ if the point (x, y) were to move along the _____ line to the curve $y = f(x)$ instead of moving along the _____ itself. This is expressed by the approximation

$$f'(x) \cdot \Delta x \approx \underline{\qquad} .$$

58.   Using the differentiation process for the function $f(x) = ax^3 + bx^2 + cx + d$, where a, b, c, d are constants, we obtain the following:

STEP 1. $f(x + \Delta x) =$ _____ .

STEP 2. $f(x + \Delta x) - f(x) =$ _____ .

STEP 3. $\dfrac{f(x + \Delta x) - f(x)}{\Delta x} =$ _____ .

STEP 4. $f'(x) = \lim\limits_{\Delta x \to 0} \dfrac{f(x + \Delta x) - f(x)}{\Delta x} =$ _____ .

We will use this result in the following application.

59.   Suppose the law of motion of a particle is given by

$$s = t^3 - 6t^2 + 2 .$$

Then the instantaneous velocity is given by

$$v = \frac{ds}{dt} = \underline{\qquad\qquad} .$$

When t = 2.3 sec, the velocity of the particle is $v(2.3) =$ _____ . If our coordinate axis of motion is such that the positive direction is to the right (which is conventional), the interpretation of this negative velocity means that the particle is moving to the _____ . When t = 4 sec, the velocity of the particle is _____ and the particle is at rest. When t = 4.5 sec, the velocity of the particle is _____ and the particle is moving to the _____ .

---

56. $\dfrac{f(x + \Delta x) - f(x)}{\Delta x}$, average, x, $f'(x)$, instantaneous, y, x   57. y, tangent, curve, Δy

58. $a[x^3 + 3x^2\Delta x + 3x(\Delta x)^2 + (\Delta x)^3] + b[x^2 + 2x\Delta x + (\Delta x)^2] + c(x + \Delta x) + d$,

   $a[3x^2\Delta x + 3x(\Delta x)^2 + (\Delta x)^3] + b[2x\Delta x + (\Delta x)^2] + c\Delta x$,

   $a[3x^2 + 3x\Delta x + (\Delta x)^2] + b[2x + \Delta x] + c$, $3ax^2 + 2bx + c$

59. $3t^2 - 12t$, -11.73 units/sec., left, 0, $\dfrac{27}{4}$ units/sec., right

60.   Consider the equilateral triangle pictured to the right. By the Pythagorean theorem $s^2 = $ _____ or, solving for h, h = _____ . Then, the area of the triangle is given by

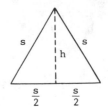

$$A = \tfrac{1}{2} \text{ base} \cdot \text{height} = \underline{\hspace{2cm}} \ .$$

The average rate of change of area with respect to side length is

$$\frac{\Delta A}{\Delta s} = \frac{\overline{\hspace{4cm}}}{\Delta s} = \underline{\hspace{3cm}} \ .$$

Taking the limit as Δs tends to zero gives,

$$\frac{dA}{ds} = \lim_{\Delta s \to 0} \frac{\Delta A}{\Delta s} = \underline{\hspace{3cm}} \ ,$$

the _____ rate of change of area with respect to _____ length for an equilateral triangle.

61.   Suppose it costs $C(x)$ thousand dollars per year to produce x thousand gallons of antifreeze, where $C(x)$ is given by the table

x	.25	.5	.75	1.0	1.25	1.50	1.75	2.0	2.25	2.5
C(x)	5.875	8.5	10.875	13.0	14.875	16.5	17.875	19.0	19.875	20.0

The _____ cost at any x is the value of the derivative $C'(x)$. Using the table, we estimate $C'(1.75)$ as follows:

$$C'(1.75) \approx \frac{\Delta C}{\Delta x} = \frac{\overline{\hspace{3cm}}}{2.0 - 1.75} = \underline{\hspace{3cm}} \ .$$

Here we have estimated the marginal cost by the _____ cost.

1-9   LIMITS.

This article is of a more technical nature than the preceding one. It is intended to make the limiting concept for a function precise. The idea of limit lies at the very heart of calculus.

OBJECTIVE A :   Write the formal definition of the limit of a function $F(t)$ as t approaches a number c.

---

60. $s^2 = h^2 + \frac{s^2}{4}$, h = $\frac{\sqrt{3}}{2}$ s, $\frac{\sqrt{3}}{4}$ s^2, $\frac{\sqrt{3}}{4}$ (s + Δs)^2 - $\frac{\sqrt{3}}{4}$ s^2, $\frac{\sqrt{3}}{4}$ (2s + Δs), $\frac{\sqrt{3}}{2}$ s,

        instantaneous, side

61.   marginal, 19.0 - 17.875, 4.5, average

62. Let F be a function defined on an open interval containing the point c, except possibly at c itself. Then the limit of F as t approaches c is L, written

_____ ,

if, given any positive number ϵ, there is a positive number δ such that _____ holds whenever $0 < |t-c| < \delta$.

63. As an application of the definition, consider the limit of the function $F(t) = 3 - 2t$ as t approaches 5. The limit is $L = -7$. To show this it is required to establish that: For any positive ϵ, there is a positive number δ such that

_____ when $0 <$ _____ $< \delta$.

Now, $|(3 - 2t) - (-7)| = 2 \cdot$ _____ . Thus,

$2|t - 5| < \epsilon$ provided $|t - 5| <$ _____ .

Therefore, if $\delta =$ _____, then

$|(3 - 2t) - (-7)| < \epsilon$ whenever _____ .

That is, $\lim_{t \to 5} (3 - 2t) = -7$.

OBJECTIVE B : Evaluate limits $\lim_{t \to c} f(t)$ when $f(t)$ is a sum, product, or quotient of polynomials in t.

64. $\lim_{x \to -2} (4 + 3x)(x^2 - x + 1) = \lim_{x \to -2} (4 + 3x)$ _____

= _____ $(4 + 2 + 1) = -14$.

65. $\lim_{t \to 4} \dfrac{t^2 + t - 2}{t^2 - 1} = \dfrac{\text{_____}}{\lim_{t \to 4} (t^2 - 1)} = \dfrac{18}{\text{___}}$.

66. $\lim_{t \to 1} \dfrac{t^3 + t^2 - 3t + 1}{t - 1} \neq \dfrac{\lim_{t \to 1} (t^3 + t^2 - 3t + 1)}{\text{_____}}$

because the limit of the denominator is _____ .

However, $\dfrac{t^3 + t^2 - 3t + 1}{t - 1} = \dfrac{(t - 1)(\text{_____})}{t - 1}$,

so that $\lim_{t \to 1} \dfrac{t^3 + t^2 - 3t + 1}{t - 1} = \lim_{t \to 1}$ _____ = _____ .

---

62. $\lim_{t \to c} F(t) = L$, $|F(t) - L| < \epsilon$ 63. $|(3 - 2t) - (-7)| < \epsilon$, $|t - 5|$, $|t - 5|$, $\frac{\epsilon}{2}$, $\frac{\epsilon}{2}$, $0 < |t - 5| < \delta$

64. $\lim_{t \to -2} (x^2 - x + 1)$, -2   65. $\lim_{t \to 4} (t^2 + t - 2)$, 15

66. $\lim_{t \to 1} (t - 1)$, 0, $t^2 + 2t - 1$, $t^2 + 2t - 1$, 2

67. $\displaystyle\lim_{h\to0} \frac{(1+h)^3 - 1}{h} = \lim_{h\to0}\frac{(\underline{\hspace{2cm}}) - 1}{h}$

$= \displaystyle\lim_{h\to0} \frac{\overline{\underline{\hspace{3cm}}}}{h}$

$= \displaystyle\lim_{h\to0} \underline{\hspace{3cm}} = \underline{\hspace{2cm}}$.

OBJECTIVE C: Specify the five important limit properties of the two limit theorems stated in the text.

Assuming all limits exist and are finite:

68. $\displaystyle\lim_{t\to c}[f(t) \pm g(t)] = \underline{\hspace{4cm}}$.

69. $\displaystyle\lim_{t\to c}[k\,f(t)] = \underline{\hspace{3cm}}$ for every number k.

70. $\displaystyle\lim_{t\to c}[f(t) \cdot g(t)] = \underline{\hspace{5cm}}$.

71. $\displaystyle\lim_{t\to c}\left[\frac{f(t)}{g(t)}\right] = \underline{\hspace{4cm}}$, provided $\underline{\hspace{4cm}}$.

72. If $f(t) \le g(t) \le h(t)$ for all values of t near c, and if $\displaystyle\lim_{t\to c}f(t) = \lim_{t\to c}h(t) = L$, then $\underline{\hspace{4cm}}$.

OBJECTIVE D: For elementary functions $y = f(x)$, find the righthand and lefthand limits as x approaches c, and from these determine if $\displaystyle\lim_{x\to c}f(x)$ exists.

73. Consider the function defined by

$$f(x) = \begin{cases} 4 - x^2 & \text{if } x \le -1 \\ 1 + x^2 & \text{if } x > -1 \end{cases}$$

The graph is shown at the right.

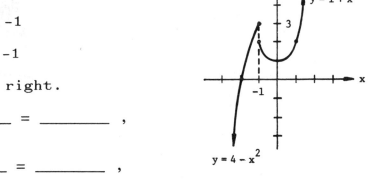

$\displaystyle\lim_{x\to-1^+}f(x) = \lim_{x\to-1^+}\underline{\hspace{2cm}} = \underline{\hspace{1.5cm}}$,

$\displaystyle\lim_{x\to-1^-}f(x) = \lim_{x\to-1^-}\underline{\hspace{2cm}} = \underline{\hspace{1.5cm}}$,

Since $\displaystyle\lim_{x\to-1^+}f(x) \ne \lim_{x\to-1^-}f(x)$, the limit $\displaystyle\lim_{x\to-1}f(x)$ $\underline{\hspace{1.5cm}}$ exist.

---

67. $1 + 3h + 3h^2 + h^3,\ 3h + 3h^2 + h^3,\ 3 + 3h + h^2,\ 3$ 68. $\displaystyle\lim_{t\to c}f(t) \pm \lim_{t\to c}g(t)$

69. $k\,\displaystyle\lim_{t\to c}f(t)$ 70. $\displaystyle\lim_{t\to c}f(t) \cdot \lim_{t\to c}g(t)$ 71. $\dfrac{\displaystyle\lim_{t\to c}f(t)}{\displaystyle\lim_{t\to c}g(t)},\ \displaystyle\lim_{t\to c}g(t) \ne 0$

72. $\displaystyle\lim_{t\to c}g(t) = L$ 73. $1 + x^2,\ 2,\ 4 - x^2,\ 3,$ does not

OBJECTIVE E : Evaluate limits of trigonometric functions by making use of appropriate trigonometric identities and the limit theorems.

74. $\lim\limits_{x\to 0} \dfrac{1 - \cos x}{1 - \sin x} = \dfrac{\lim\limits_{x\to 0} (1 - \cos x)}{\underline{\hspace{2cm}}} = \dfrac{(1 - \underline{\hspace{1cm}})}{(1 - 0)} = \underline{\hspace{2cm}}$.

75. $\lim\limits_{x\to 0} \dfrac{h}{\sin h} = \lim\limits_{h\to 0} \dfrac{1}{\underline{\hspace{1.5cm}}} = \dfrac{1}{\lim\limits_{h\to 0} \dfrac{\sin h}{\underline{\hspace{1cm}}}} = \dfrac{1}{\underline{\hspace{0.8cm}}} = \underline{\hspace{2cm}}$.

76. $\lim\limits_{x\to 0} \left(\dfrac{x + 1}{2x}\right) \sin x = \lim\limits_{x\to 0} \dfrac{1}{2}(x + 1) \cdot \lim\limits_{x\to 0} \underline{\hspace{2.5cm}}$

$= \underline{\hspace{2.5cm}} \cdot 1 = \underline{\hspace{2cm}}$.

1-10   INFINITY AS A LIMIT.

OBJECTIVE : Calculate the limit of $f(x)$ as x approaches $+\infty$ or $-\infty$, whenever the limit exists.

77. $\lim\limits_{x\to\infty} \dfrac{5x^3}{1 + 3x - 2x^3} = \lim\limits_{h\to 0} \dfrac{\dfrac{5}{h^3}}{\underline{\hspace{1.5cm}}} = \lim\limits_{h\to 0} \dfrac{5}{\underline{\hspace{1.5cm}}} = \underline{\hspace{2cm}}$.

78. $\lim\limits_{t\to-\infty} \dfrac{1 - \cos t}{t} = \lim\limits_{t\to-\infty} \underline{\hspace{1.5cm}} \cdot \lim\limits_{t\to-\infty} (1 - \cos t) \le \lim\limits_{t\to-\infty} \dfrac{1}{t} \cdot \underline{\hspace{1.5cm}}$

$= \underline{\hspace{2cm}}$.

Also, $\lim\limits_{t\to-\infty} \dfrac{1 - \cos t}{t} = \lim\limits_{t\to-\infty} \dfrac{1}{t} \cdot \underline{\hspace{1.5cm}} \ge \lim\limits_{t\to-\infty} \dfrac{1}{t} \cdot \underline{\hspace{1.5cm}}$

$= \underline{\hspace{2cm}}$.

Therefore, by the Sandwich Theorem, $\lim\limits_{t\to-\infty} \dfrac{1 - \cos t}{t} - \underline{\hspace{2cm}}$.

1-11   CONTINUITY.

OBJECTIVE A : Define precisely what is meant for a function f to be continuous at the point $x = c$.

79. The three conditions that must be satisfied if the function f is to be continuous at the point $x = c$ are that $\underline{\hspace{2cm}}$ exists, $\underline{\hspace{2.5cm}}$ exists, and $\underline{\hspace{3cm}}$.

---

74. $\lim\limits_{x\to 0} (1 - \sin x)$, 1, 0   75. $\dfrac{\sin h}{h}$, h, 1, 1   76. $\dfrac{\sin x}{x}$, $\dfrac{1}{2}$, $\dfrac{1}{2}$

77. $1 + \dfrac{3}{h} - \dfrac{2}{h^3}$, $h^3 + 3h^2 - 2$, $-\dfrac{5}{2}$ 78. $\dfrac{1}{t}$, 2, 0, $\lim\limits_{t\to-\infty} (1 - \cos t)$, -2, 0, 0

79. $f(c)$, $\lim\limits_{x\to c} f(x)$, $\lim\limits_{x\to c} f(x) = f(c)$

80. A function is continuous over an interval if it is continuous at _____ within that interval.

81. If a function f is not continuous at the point x = c, it is said to be _____ at c.

OBJECTIVE B : Given an elementary function y = f(x), determine its points of continuity and discontinuity. Be able to justify your conclusions.

82. Consider $f(x) = \begin{cases} x + 4 & \text{if } x < -1 \\ -x & \text{if } x \geq -1 \end{cases}$. Observe that c = -1 belongs to the domain of f: f(-1) = 1. Does f have a limit as x → -1? To answer that question, we calculate the right and lefthand limits:

$$\lim_{x \to -1^-} f(x) = \lim_{x \to -1^-} (\underline{\quad\quad}) = \underline{\quad\quad} \ ,$$

$$\lim_{x \to -1^+} f(x) = \lim_{x \to -1^+} (\underline{\quad\quad}) = \underline{\quad\quad} \ .$$

Since $\lim_{x \to -1^-} f(x) \neq \lim_{x \to -1^+} f(x)$, then $\lim_{x \to -1} f(x)$ _____ .

83. Let $f(x) = \dfrac{x}{x - 1}$. Since x = _____ does not belong to the domain of f we conclude that f is _____ at 1. Also, $\lim_{x \to 1^-} f(x) =$ _____ and $\lim_{x \to 1^+} f(x) =$ _____ so f does not have a finite limit as x → 1. However, as x → +∞ or x → -∞, f(x) → _____ . Sketch a graph of f. Observe that f is continuous at all points except x = 1.

OBJECTIVE C : Specify the main facts related to continuous functions.

84. If f and g are continuous at c, then f + g, f - g, and f · g are _____ at c.

85. If f and g are continuous at c, then $\frac{f}{g}$ is _____ at c provided that _____ .

---

80. all points

81. discontinuous

82. x + 4, 3,
    -x, 1,
    does not exist,
    discontinuous

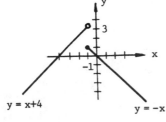

83. 1,
    discontinuous,
    -∞, +∞, 1

84. continuous

85. continuous, g(c) ≠ 0

86. If f is continuous at c, and k is any constant, then kf is _____ at c.

87. Every constant function is continuous _____ .

88. Every polynomial function is continuous _____ .

89. Every rational function is continuous _____ _____ .

90. If f is differentiable at x = c, then _____ _____ .

91. Suppose that f(x) is continuous for all x in the closed interval [a,b]. Then f has a _____ value m and a _____ value M on [a,b].

92. If f is continuous at c and g is continuous at f(c), then the composite _____ is continuous at _____ .

93. Suppose that f(x) is continuous for all x in the closed interval [a,b], and that N is any number between f(a) and f(b). What is your conclusion? _____ _____ .

---

86. continuous   87. at every number   88. at every number

89. at every number at which the denominator is not zero

90. f is continuous at $x = c$ 91. minimum, maximum

92. $g(f(x))$, $x = c$ 93. There is at least one number c between a and b such that $f(c) = N$.

CHAPTER 1   SELF-TEST

1. For each of the following, draw a pair of coordinate axes, plot the given point, and plot the point meeting the specified requirement giving the coordinates of the second point:

    (a) $P(-3,1)$, and $Q(x,y)$ so that PQ is parallel to the x-axis and bisected by the y-axis;

    (b) $P(2,-2)$, and $R(x,y)$ so that PR is perpendicular to the x-axis and bisected by it;

    (c) $P(-1.3,-0.5)$, and $S(x,y)$ so that PS is bisected by the origin.

2. A particle moves in the plane along a straight line from $P(-3,-1)$ to $Q(7,-3)$. Find the increments Δx and Δy and distance from P to Q.

3. A particle moves from the point $A(2,-3)$ to the x-axis in such a way that $\Delta y = -6\Delta x$. What are its new coordinates?

4. Determine if the points $A(1,-3)$, $B(-2,9)$, and $C(5,-19)$ are collinear.

5. Find the slope of the line through the points $(1,4)$ and $(-3,2)$, and write an equation of the line.

6. Determine the slope, the x-intercept, and the y-intercept for each of the following equations:
    (a) $3x + 4y = -1$ (b) $x = 2$ (c) $y = -1$ (d) $x^2 = 2y - 1$

7. Find an equation of the line through the point $(5,-7)$ and perpendicular to the line $2y - x = 8$.

8. Let f be defined by the equation $f(x) = x^2 + 3x - 2$. Find the domain and range of f. Also find the values $f(-2)$, $f(-1)$, $f(0)$, $f(2)$, $f(2b)$ and $f(a + b)$, and sketch the graph of f.

9. Find the domain and range of the function $f(x) = \dfrac{x^2 - x - 6}{x + 2}$, and sketch the graph.

10. Describe the domain of the following absolute value inequalities without using absolute value symbols.
    (a) $|2x - 3| \le 5$ (b) $|x - 2| > 4$ (c) $|2 - 3x| < -1$

11. Let $f(x) = x^2 + 1$ and $g(x) = (x + 1)^2$. Find

    (a) $f(x) - g(x)$ (b) $\dfrac{f(x)}{g(x)}$ (c) $f(g(x))$

    (d) $g(f(x))$ (e) $g(g(x))$ (f) $g(x^2)$

12. Find two functions, f and g, that will produce the given composite h(x) = f(g(x)). There is no unique answer.

   (a) $h(x) = \sqrt{x^2 - 1}$

   (b) $h(t) = (t + \frac{1}{t})^5$

13. Use the increment method to find the slope of the curve $y = x^3 - 2x + 5$ at a point (x,y) on the curve.

14. Write an equation of the tangent line to the curve in Problem 13 at the point when x = -2.

15. Using the fact that for any constant k, if $f(x) = kx^3$, then the derivative $f'(x) = 2kx^2$ holds, calculate the instantaneous rate of change of the volume of a sphere with respect to its radius when the radius is 3 cm.

16. Using the fact that for constants a, b, c and quadratic function $f(t) = at^2 + bt + c$, then $f'(t) = 2at + b$, find the instantaneous speed of a particle moving along a straight line according to the equation $s = 5t^2 - 3t$ when t = 2 sec.

17. Evaluate the following limits:

   (a) $\lim_{t \to 3} \dfrac{t^2 - 1}{t - 1}$

   (b) $\lim_{x \to 2} \dfrac{2x^2 - 3x - 2}{x - 2}$

   (c) $\lim_{t \to \infty} \dfrac{t^2}{4 - t^2}$

   (d) $\lim_{x \to 1} \dfrac{3x - 1}{5x^3 - 2x + 1}$

   (e) $\lim_{x \to 0} \dfrac{\sin x^{1/3}}{x^{1/3}}$

   (f) $\lim_{x \to \frac{\pi}{2}} \dfrac{\cos x}{\pi - 2x}$

18. Let f be defined by $f(x) = \begin{cases} 2x - 3, & \text{if } x \geq 0 \\ -1, & \text{if } x < 0 \end{cases}$

   (a) Find $\lim_{x \to 0^+} f(x)$ and $\lim_{x \to 0^-} f(x)$.

   (b) Is f continuous at x = 0? Justify your answer.

19. Consider the function $f(x) = \dfrac{x - 1}{x^2 - x}$.

   (a) For what values of x is f continuous? Justify your conclusion.

   (b) Is f continuous at x = 1? If not what value can be assigned to f(1) so that the resultant function is continuous there?

SOLUTIONS TO CHAPTER 1 SELF-TEST

1.

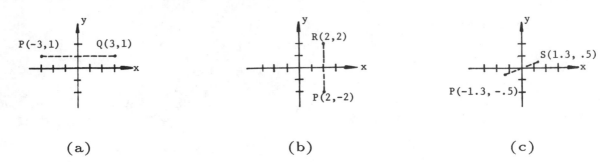

(a)   (b)   (c)

2. $\Delta x = 7 - (-3) = 10$, $\Delta y = -3 - (-1) = -2$,

$d = \sqrt{(\Delta x)^2 + (\Delta y)^2} = \sqrt{104} \approx 10.198$

3.   The new coordinates can be written as $(x,0)$ since the point lies on the x-axis.   From $\Delta y = -6\Delta x$, we have $0 - (-3) = -6(x - 2)$ or, solving, $x = 3/2$.   Thus $(\frac{3}{2},0)$ gives the coordinates of the new position of the particle.

4.   The slope of AB is $m_1 = \dfrac{9 - (-3)}{-2 - 1} = -4$ and the slope of BC is $m_2 = \dfrac{-19 - 9}{5 - (-2)} = -4$.   Since these slopes are equal, the three points do lie on a common straight line.

5. $m = \dfrac{2 - 4}{-3 - 1} = \frac{1}{2}$ is the slope, and $y - 4 = \frac{1}{2}(x - 1)$ or $2y - x = 7$ is an equation of the line.

6.   (a)   Solving algebraically for y, $y = -\frac{3}{4}x - \frac{1}{4}$.   Thus, the slope is $m = -\frac{3}{4}$, the y-intercept is $b = -\frac{1}{4}$; and when $y = 0$, $x = -\frac{1}{3}$ is the x-intercept.

   (b)   This is a vertical line so it has no slope and no y-intercept.   The x-intercept is 2.

   (c)   This is a horizontal line.   It has slope 0 and y-intercept -1.   It has no x-intercept.

   (d)   Since the variable x is squared, this equation does not represent a straight line.   When $x = 0$, $y = \frac{1}{2}$ so the y-intercept is $\frac{1}{2}$.   If $y = 0$, $x^2 = -1$ which is impossible, so it has no x-intercept.

7.   The slope of $2y - x = 8$ or $y = \frac{1}{2}x + 4$ is $m = \frac{1}{2}$.   Therefore, the slope of the perpendicular line is $m' = -2$ and an equation is given by $y + 7 = -2(x - 5)$ or $y = -2x + 3$.

8. The function $f(x) = x^2 + 3x - 2$ is defined for all values of   x,
    so the domain is $-\infty < x < \infty$ (all real numbers). Setting
    $y = x^2 + 3x - 2$ or $y + 2 = x^2 + 3x$ and completing the square on
    the righthand side gives

    $$y + 2 + \frac{9}{4} = \left(x + \frac{3}{2}\right)^2 \,, \quad \text{or} \quad y + \frac{17}{4} = \left(x + \frac{3}{2}\right)^2 \,.$$

    Thus, $y \geq -\frac{17}{4}$ so the range of f is the interval $[-\frac{17}{4}, \infty)$.

    $f(-2) = -4$, $f(-1) = -4$,

    $f(0) = -2$, $f(2) = 8$,

    $f(2b) = 4b^2 + 6b - 2$,

    $f(a + b) = a^2 + 2ab + b^2 + 3(a + b) - 2$

    The graph of f is shown at the right.

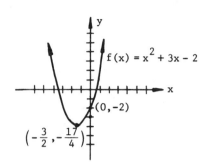

9. $\dfrac{x^2 - x - 6}{x + 2} = \dfrac{(x - 3)(x + 2)}{x + 2}$.

    Thus, the domain of f is all
    real numbers except $x = -2$.
    Also, for $x \neq -2$, $f(x) = x - 3$.
    This is a straight line with the
    point $(-2,-5)$ deleted so the
    range of f is all real numbers
    except $y = -5$.

    The graph of f is shown at the
    right.

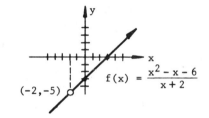

10. (a) $-5 \leq 2x - 3 \leq 5$ so $-1 \leq x \leq 4$

     (b) $x - 2 > 4$ or $-(x - 2) > 4$

           so $x > 6$ or $-x > 2$;

           that is, x lies within the union of the intervals
           $(-\infty,-2) \cup (6,\infty)$.

     (c)   Since for any real number its absolute value is nonnegative,
           it is impossible for $|2 - 3x|$ to be less than -1;   thus
           the domain of x is empty (there is no solution).

11. (a) $f(x) - g(x) = -2x$ (b) $\dfrac{f(x)}{g(x)} = \dfrac{x^2 + 1}{(x + 1)^2}$

     (c) $f(g(x)) = (x + 1)^4 + 1$ (d) $g(f(x)) = (x^2 + 2)^2$

     (e) $g(g(x)) = (x^2 + 2x + 2)^2$ (f) $g(x^2) = (x^2 + 1^2)$

12.   (a) $f(x) = \sqrt{x}$ and $g(x) = x^2 - 1$

(b) $f(t) = t^5$ and $g(t) = t + \frac{1}{t}$

13.   STEP 1. $y + \Delta y = (x + \Delta x)^3 - 2(x + \Delta x) + 5$
$$= x^3 + 3x^2\Delta x + 3x(\Delta x)^2 + (\Delta x)^3 - 2x - 2\Delta x + 5$$

STEP 2.   Subtracting $(y + \Delta y) - y$:
$$\Delta y = 3x^2\Delta x + 3x(\Delta x)^2 + (\Delta x)^3 - 2\Delta x$$

STEP 3.   Dividing by Δx yields,
$$m_{sec} = \frac{\Delta y}{\Delta x} = 3x^2 + 3x\Delta x + (\Delta x)^2 - 2$$

STEP 4.   As Δx tends to zero
$$m_{tan} = \lim_{\Delta x \to 0} \frac{\Delta y}{\Delta x} = 3x^2 - 2 \ .$$

14.   When $x = -2$, $y = (-2)^3 - 2(-2) + 5 = 1$,   and from the preceding problem solution, $m = 3(-2)^2 - 2 = 10$.   Thus, $y - 1 = 10(x + 2)$ or $y = 10x + 21$ is an equation of the tangent line.

15.   The volume of a sphere is given by $V = \frac{4}{3}\pi r^3$,   where r is the radius.   We seek the value of $\frac{dV}{dr}$ when $r = 3$.   Thus, from the derivative formula, $\frac{dV}{dr} = V'(r) = 4\pi r^2$ so that $V'(3) = 36\pi$.

16.   We seek the velocity $v = \frac{ds}{dt}$ when $t = 2$.   From the given derivative formula $v = \frac{ds}{dt} = 10t - 3$ so that $v(2) = 17$ units/sec.

17.   (a) $\lim\limits_{t \to 3} \dfrac{t^2 - 1}{t - 1} = \dfrac{9 - 1}{3 - 1} = 4$

(b) $\lim\limits_{x \to 2} \dfrac{2x^2 - 3x - 2}{x - 2} = \lim\limits_{x \to 2} \dfrac{(2x + 1)(x - 2)}{x - 2} = \lim\limits_{x \to 2} 2x + 1 = 5$

(c) $\lim\limits_{t \to \infty} \dfrac{t^2}{4 - t^2} = \lim\limits_{h \to 0} \dfrac{\frac{1}{h^2}}{4 - \frac{1}{h^2}} = \lim\limits_{h \to 0} \dfrac{1}{4h^2 - 1} = \dfrac{1}{0 - 1} = -1$

(d) $\lim\limits_{x \to 1} \dfrac{3x - 1}{5x^3 - 2x + 1} = \dfrac{3 - 1}{5 - 2 + 1} = \dfrac{1}{2}$

(e)   Let $u = x^{1/3}$ so that $x \to 0$ is equivalent to $u \to 0$.

Thus $\lim\limits_{x \to 0} \dfrac{\sin x^{1/3}}{x^{1/3}} = \lim\limits_{u \to 0} \dfrac{\sin u}{u} = 1$.

(f) $\lim\limits_{x \to \frac{\pi}{2}} \dfrac{\cos x}{\pi - 2x} = \lim\limits_{x \to \frac{\pi}{2}} \dfrac{\frac{1}{2}\cos x}{\frac{\pi}{2} - x}$;   let $u = \frac{\pi}{2} - x$ so that $x \to \frac{\pi}{2}$

is equivalent to $u \to 0$.   Thus,

$$\lim_{x \to \frac{\pi}{2}} \frac{\frac{1}{2} \cos x}{\frac{\pi}{2} - x} = \lim_{u \to 0} \frac{\frac{1}{2} \cos\left(\frac{\pi}{2} - u\right)}{u} = \lim_{u \to 0} \frac{1}{2} \frac{\sin u}{u} = \frac{1}{2} \, .$$

18. (a) From the graph of f shown
       at the right,

       $$\lim_{x \to 0^+} f(x) = \lim_{x \to 0^+} (2x - 3) = -3$$

       $$\lim_{x \to 0^-} f(x) = \lim_{x \to 0^-} (-1) = -1$$

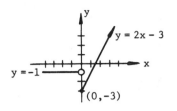

    (b) No, $\lim_{x \to 0} f(x)$ does not exist because the lefthand and
       righthand limits differ as x tends to zero, so f is not
       continuous at x = 0.

19. (a) Since division by zero is never permitted, the points x = 0
       and x = 1 do not belong to the domain of f, and
       therefore f is discontinuous at those two values; it is
       continuous for all other values of x.

    (b) $f(x) = \dfrac{x - 1}{x^2 - x} = \dfrac{x - 1}{x(x - 1)} = \dfrac{1}{x}$ if $x \neq 1$ and $x \neq 0$. Since

       $$\lim_{x \to 1} f(x) = \lim_{x \to 1} \frac{1}{x} = 1 \, ,$$ if we specify f(1) = 1 the new

       function so defined is continuous at x = 1.

NOTES.

CHAPTER 2 DERIVATIVES

2-1 POLYNOMIAL FUNCTIONS AND THEIR DERIVATIVES.

1. If c is a constant and $y = f(x) = c$, then $\frac{dy}{dx} =$ ___0___ .

2. If n is any positive integer and $y = f(x) = x^n$, then

$$\frac{dy}{dx} = \underline{nx^{(n-1)}} \ .$$

3. If $u = f(x)$ is a differentiable function of x, and if $y = cu$ where c is a constant, then

$$\frac{dy}{dx} = \underline{cf'(x)} \ .$$

4. If $u = f(x)$ and $v = g(x)$ are differentiable functions of x, then $y = u + v$ is a __differentiable__ function of x, and

$$\frac{dy}{dx} = \underline{f'(x) + g'(x)} = \underline{dy/dx + dv/dx} \ .$$

OBJECTIVE A : Calculate the derivative of any polynomial function.

5. $\frac{d}{dx}(3x^2 - 12x + 1) = \underline{6x - 12}$.

6. $\frac{d}{dx}\left(\sqrt{3}x^4 - \frac{2}{5}x^3 + \frac{1}{3}x^2 - 15x + 109\right) = \underline{4\sqrt{3}x^3 - \frac{6}{5}x^2 + \frac{2}{3}x - 15}$

$$4\sqrt{3}x$$

OBJECTIVE B : Calculate second and higher order derivatives of any polynomial function.

7. If $y = f(x)$, the second derivative of y with respect to x is the derivative of __$f'(x)$__ . The second derivative is denoted by __$f''(x)$__ or __y''__ or __d^2y/dx^2__ .

8. In general, the nth derivative of $y = f(x)$ with respect to x is the derivative of __$d^{n-1}y/dx^{n-1}$__ , and is denoted by __d^ny/dx^n__ or __y^n__ or __$f^n(x)$__ .

---

1. 0

2. nx^{n-1}

3. $c\frac{du}{dx} = cf'(x)$

4. differentiable, $\frac{du}{dx} + \frac{dv}{dx} = f'(x) + g'(x)$

5. $6x - 12$

6. $4\sqrt{3}\ x^3 - \frac{6}{5}\ x^2 + \frac{2}{3}\ x - 15$

7. $\frac{dy}{dx} = f'(x), \ \frac{d^2y}{dx^2}, \ y'', \ \text{or} \ f''(x)$

8. $\frac{d^{n-1}y}{dx^{n-1}}, \ \frac{d^ny}{dx^n}, \ y^{(n)}, \ \text{or} \ f^{(n)}(x)$

9. $\frac{d^2}{dx^2}(4x^5 - 3x^2 + 2x - 20) = \frac{d}{dx}\left(20x^4 - 6x + 2\right) = \underline{80x^3 - 6}$.

10. $\frac{d^2}{dx^2}(2x^2 - 1)(x - 3) = \frac{d^2}{dx^2}\left(2x^3 - 6x^2 - x + 3\right) = \frac{d}{dx}\left(6x^2 - 12x - 1\right)$
    $= \underline{12x - 12}$.

OBJECTIVE C : If s = f(t) gives the position of a moving body as a
function of time t, find and interpret the velocity
and acceleration at a specified instant.

11. Suppose a particle is moving along a straight line, negative to
the left and positive to the right, according to the law

$$s = t^3 - 3t^2 - 9t + 5 .$$

Then the velocity is given by $\frac{ds}{dt} = \underline{3t^2 - 6t - 9 = 3(t^2 - 2 - 3)}$.
Thus, the velocity is positive when $\underline{t < 0 - 1}$ or $\underline{t > 3}$ and
the particle is moving to the _____ ; the velocity is
_____ when -1 < t < 3 so the particle is moving to the
_____ .

12. The acceleration of the particle is $\frac{d^2s}{dt^2} = $ _____ . When
the velocity is zero, t = _____ or _____ and the
acceleration has the value _____ or _____, respectively.

13. Suppose a ball is thrown directly upward with a speed of
96 ft/sec and moves according to the law $y = 96t - 16t^2$, where
y is the height in feet above the starting point, and t is
the time in seconds after it is thrown.   The velocity of the
ball at any time t is $v(t) = \frac{dy}{dt} = $ _____ . Hence when
t = 2 sec, the velocity of the ball is _____ . Since
v(2) is positive, the ball is still rising. At its highest
point the velocity of the ball is _____ , and this occurs
when t = _____ seconds.   The height corresponding to this
time is y = _____ feet, and this is the highest point
reached.   Notice the acceleration is a constant _____ ft/sec^2.

OBJECTIVE D : Find an equation of the tangent line to a curve
y = f(x) meeting some specified requirement (such as
a condition on the slope).

14. Consider the curve $y = x^3 - 9x^2 + 15x - 5$.   The derivative y'
gives the value of the _____ of the tangent line at any x.
For this particular curve, $y' = $ _____ $= 3(\underline{})(x - 5)$.

---

9. $20x^4 - 6x + 2$, $80x^3 - 6$ 10. $2x^3 - 6x^2 - x + 3$, $6x^2 - 12x - 1$, $12x - 12$

11. $3t^2 - 6t - 9 = 3(t - 3)(t + 1)$, $t < -1$ or $t > 3$, right, negative, left

12. $6t - 6$, -1 or 3, -12 or 12   13. $96 - 32t$, 32 ft/sec, zero, 3, $y = 144$, -32

14. slope, $3x^2 - 18x + 15$, $x - 1$, 1, 5

Thus, the tangent line is parallel to the x-axis when
x = _____ or x = _____ .

15. When the slope of the tangent line to the above curve equals
15 the value of x is _____ or _____ . The corresponding
y values are _____ and _____ , respectively. Equations of
the two tangent lines are then given by _____ and
_____ .

2-2 PRODUCTS, POWERS, AND QUOTIENTS.

OBJECTIVE A : Find the derivative of a product of polynomial
functions.

16. If $u = f(x)$ and $v = g(x)$ are differentiable functions of
x, then the derivative $\frac{d}{dx}(uv) =$ _____ .

17. If $y = (x^2 - 2)(2x^3 - 5)$, then

$y' = (x^2 - 2) \frac{d}{dx}(2x^3 - 5) + (2x^3 - 5)$ _____

$= (x^2 - 2)(_____) + (2x^3 - 5)(2x)$

$=$ _____ $+ 4x^4 - 10x =$ _____ .

18. $\frac{d}{dx}[(3x^2 - 2x + 1)(5x - 4)]$

$= (3x^2 - 2x + 1)(_____) + (5x - 4)(_____)$

$= (15x^2 - 10x + 5) + (_____) =$ _____ .

OBJECTIVE B : Find the derivative of a quotient of polynomial
functions.

19. A _____ function is a quotient of two polynomials and
is generally not itself a polynomial. Every polynomial may be
considered to be a rational function with denominator the
constant polynomial 1.

20. If $u = f(x)$ and $v = g(x)$ are differentiable functions of
x, then the derivative $\frac{d}{dx}\left(\frac{u}{v}\right) =$ _____ when $v \neq 0$.

---

15. 0, 6, -5, -23, $y + 5 = 15x$ and $y + 23 = 15(x - 6)$

16. $u\frac{dv}{dx} + v\frac{du}{dx} = f(x)g'(x) + g(x)f'(x)$ 17. $\frac{d}{dx}(x^2 - 2)$, $6x^2$, $6x^4 - 12x^2$, $10x^4 - 12x^2 - 10x$

18. 5, $6x - 2$, $30x^2 - 34x + 8$, $45x^2 - 44x + 13$

19. rational   20. $\frac{v\frac{du}{dx} - u\frac{dv}{dx}}{v^2} = \frac{g(x)f'(x) - f(x)g'(x)}{[g(x)]^2}$

21.   If $y = \dfrac{3x}{5x^2 - 1}$,   then

$$y' = \frac{(5x^2 - 1)(\underline{\hspace{1.2cm}}) - (3x)(\underline{\hspace{1.2cm}})}{(5x^2 - 1)^2} = \frac{\underline{\hspace{2cm}}}{(5x^2 - 1)^2} \ .$$

22. $\dfrac{d}{dx}\left(\dfrac{1}{x}\right) = \dfrac{x(\underline{\hspace{1cm}}) - 1(\underline{\hspace{1cm}})}{x^2} = \underline{\hspace{2.5cm}} \ .$

23. $\dfrac{d}{dt}\left(\dfrac{t^2 - 2t + 5}{t^3 + 1}\right) = \dfrac{(t^3 + 1)(\underline{\hspace{1.5cm}}) - (\underline{\hspace{1.5cm}})(3t^2)}{(t^3 + 1)^2}$

$$= \frac{2(\underline{\hspace{1.5cm}}) - 3t^4 + 6t^3 - 15t^2}{(t^3 + 1)^2}$$

$$= \frac{(\underline{\hspace{1.5cm}})}{(t^3 + 1)^2}$$

OBJECTIVE C :   Find the derivative of an integral power of a differentiable function.

24.   If $u = g(x)$ is a differentiable function of x and n is a positive integer, then the derivative

$$\frac{d}{dx}(u^n) = \underline{\hspace{2.5cm}} \ .$$

25.   The above power rule holds when n is a negative integer at all points x where $g(x)$ is $\underline{\hspace{2cm}}$.

26.   If $y = (5x^3 - x^2 + 7)^4$,   then

$$y' = 4(5x^3 - x^2 + 7)^3 \frac{d}{dx}(\underline{\hspace{2cm}}) = \underline{\hspace{3cm}} \ .$$

27. $\dfrac{d}{dx}\left[(2x - 1)\right]^{-3} = -3(\underline{\hspace{1.2cm}})^{-4} \dfrac{d}{dx}(\underline{\hspace{1.5cm}})$

$$= \underline{\hspace{2.5cm}} \ , \quad \text{if}\ \ x \neq \underline{\hspace{1.5cm}} \ .$$

28.   Let $y = \dfrac{2}{(x - 1)^3} + \dfrac{3}{1 - x^4}$.   Then $y = 2(x - 1)^{-3} + 3(1 - x^4)^{-1}$,

so that $\dfrac{dy}{dx} = -6(\underline{\hspace{2cm}}) \dfrac{d}{dx}(x - 1) - 3(1 - x^4)^{-2} \dfrac{d}{dx}(\underline{\hspace{1.5cm}})$

$$= \underline{\hspace{3cm}} \ .$$

---

21. 3, 10x, $-15x^2 - 3$ 22. 0, 1, $-\dfrac{1}{x^2}$

23. $2t - 2$, $t^2 - 2t + 5$, $t^4 - t^3 + t - 1$, $-t^4 + 4t^3 - 15t^2 + 2t - 2$

24. $nu^{n-1} \dfrac{du}{dx}$ 25. not zero

26. $5x^3 - x^2 + 7$, $4(5x^3 - x^2 + 7)^3 (15x^2 - 2x)$ 27. $2x - 1$, $2x - 1$, $-6(2x - 1)^{-4}$, $\dfrac{1}{2}$

28. $(x - 1)^{-4}$, $1 - x^4$, $-6(x - 1)^{-4} + 12x^3(1 - x^4)^{-2}$

29. $\dfrac{d}{dx} \left(\dfrac{x - 2}{x + 1}\right)^5 = 5(\underline{\hspace{2cm}})^4 \dfrac{d}{dx} (\underline{\hspace{2cm}})$

$= 5 \left(\dfrac{x - 2}{x + 1}\right)^4 \dfrac{(x + 1)(1) - (\underline{\hspace{1.5cm}})(1)}{(x + 1)^2} = \dfrac{\overline{\hspace{2cm}}}{(x + 1)^6} \; .$

2-3   IMPLICIT DIFFERENTIATION AND FRACTIONAL POWERS.

OBJECTIVE A :   Compute first and second derivatives by the technique
of implicit differentiation.

30.   An equation involving the variables   x   and   y   is said to
determine   y   _____   as a function of   x,   say   y = f(x),
provided that   f   satisfies the equation.

31.   For instance, consider the equation $x^2 + y^2 = 2$.   If we
substitute $y = \sqrt{2 - x^2}$ into the equation we obtain
$$x^2 + \left(\sqrt{2 - x^2}\right)^2 = x^2 + (\underline{\hspace{1cm}}) = \underline{\hspace{1.5cm}} \; ,$$
so the equation is satisfied.   Similarly, if we substitute
$y = - \sqrt{2 - x^2}$ into the equation we obtain
$$x^2 + \left(- \sqrt{2 - x^2}\right)^2 = x^2 + (\underline{\hspace{1.5cm}}) = \underline{\hspace{1cm}} \; ,$$
and the equation is again satisfied.   Therefore, each of the
two functions $y = \sqrt{2 - x^2}$ and $y = - \sqrt{2 - x^2}$ is defined
_____   by the equation $x^2 + y^2 = 2$.

32.   To calculate the derivative   dy/dx   for $x^2 + y^2 = 2$,
differentiate both sides of the equation with respect to   x
and solve for   dy/dx:

$$\dfrac{d}{dx} (x^2 + y^2) = \dfrac{d}{dx} (2) \quad \text{or} \quad \dfrac{d}{dx} (\underline{\hspace{1.5cm}}) + \dfrac{d}{dx} (y^2) = \dfrac{d}{dx} (2) \; .$$

Thus, $2x + (\underline{\hspace{1.5cm}}) = 0$ and solving for   dy/dx,
$\dfrac{dy}{dx} = \underline{\hspace{2.5cm}}$.   This derivative is valid whenever   y
satisfies the condition _____ .

33.   To calculate d^2y/dx^2 for $x^2 + y^2 = 2$,   differentiate both
sides of the derivative equation   dy/dx = -x/y   with respect to
x:

$$\dfrac{d}{dx} \left(\dfrac{dy}{dx}\right) = \dfrac{d}{dx}\left(- \dfrac{x}{y}\right) = - \dfrac{d}{dx} \left(\dfrac{x}{y}\right) \; , \quad \text{or}$$

$$\dfrac{d^2y}{dx^2} = - \left[\dfrac{(\underline{\hspace{1cm}}) \dfrac{dx}{dx} - x(\underline{\hspace{1cm}})}{y^2}\right] \; .$$

---

29. $\dfrac{x - 2}{x + 1}$, $\dfrac{x - 2}{x + 1}$, x - 2, $15(x - 2)^4$ 30. implicitly

31. $2 - x^2$, 2, $2 - x^2$, 2, implicitly   32. x^2, $2y\dfrac{dy}{dx}$, $-\dfrac{x}{y}$, y ≠ 0

Substitution of -x/y for dy/dx in the last equation gives

$$\frac{d^2y}{dx^2} = -\left[\frac{y - x(\underline{\quad})}{y^2}\right] = -\frac{\underline{\qquad}}{y^3} = \frac{\underline{\qquad}}{y^3}\ .$$

OBJECTIVE B :   Find the derivative of $g(x) = x^n$ when n is any rational number n = p/q.

34.   If $g(x) = x^n$ with n = 1/m, where m is a positive odd integer, then $g'(x) =$ _____ for x satisfying _____ .

35.   If $g(x) = x^n$ with n = 1/m, where m is a positive even integer, then $g'(x) =$ _____ for x satisfying _____ .

36. $\frac{d}{dx}(x^{1/9}) =$ _____ provided x satisfies _____ .

37. $\frac{d}{dx}(x^{1/6}) =$ _____ provided x satisfies _____ .

38. $\frac{d}{dx}(x^{3/5}) =$ _____ provided x satisfies _____ .

39. $\frac{d}{dx}(x^{-3/4}) =$ _____ provided x satisfies _____ .

40.   Let $y = \sqrt{\frac{x + 3}{x - 3}}$, so $y = u^{1/2}$ where $u = \frac{x + 3}{x - 3}$.   Then

$$\frac{dy}{dx} = \frac{1}{2} u^{-1/2} \frac{du}{dx}\ \text{whenever}\ u > 0.\ \ \text{Thus}$$

$$\frac{dy}{dx} = \frac{1}{2}\left(\underline{\qquad}\right)^{-1/2} \frac{(x - 3)(1) - (\underline{\qquad})}{(x - 3)^2}$$

$$= \frac{1}{2\sqrt{\frac{x + 3}{x - 3}}}\ \underline{\qquad}\ \text{whenever}\ \frac{x + 3}{x - 3} > 0.$$

OBJECTIVE C :   Find the lines that are tangent and normal to a specified curve at a given point.

41.   By direct substitution the point (1,1) lies on the curve xy = 1. Differentiating both sides with respect to x yields y + _____ = 0. Solving for dy/dx gives

$$\frac{dy}{dx} = \underline{\qquad}\ .$$

---

33. y, $\frac{dy}{dx}$, $-\frac{x}{y}$, $x^2 + y^2$, 2 (because $x^2 + y^2 = 2$)   34. nx^{n-1}, $x \neq 0$

35. nx^{n-1}, $x > 0$ 36. $\frac{1}{9} x^{-8/9}$, $x \neq 0$ 37. $\frac{1}{6} x^{-5/6}$, $x > 0$

38. $\frac{3}{5} x^{-2/5}$, $x \neq 0$ 39. $-\frac{3}{4} x^{-7/4}$, $x > 0$ 40. $\frac{x + 3}{x - 3}$, $(x + 3)(1)$, $\frac{-6}{(x - 3)^2}$

Thus, at   (1,1)   the slope of the curve is $\frac{dy}{dx}\big|_{(1,1)}$ = _____ .

Therefore, the tangent to the curve at the point   (1,1)   is

y - 1 = _____ .

42.   For the curve in Problem 41, the slope of the normal is _____ at the point   (1,1).   The normal is therefore given by the equation

y - 1 = _____ .

2-4   LINEAR APPROXIMATIONS AND DIFFERENTIALS.

OBJECTIVE A :   Given a function   y = f(x)   and a point   x = a,   find the linearization of   f(x)   near   a.   Use your linearization to estimate a specified function value.

43.   If   y = f(x)   is differentiable at   x = a,   then the <u>linearization</u> of   f   at   a   is given by

$$L(x) = \underline{f(a) + f'(a)(x - a)}$$

44.   To find the linearization to $f(x) = 2\frac{1}{2}x^2 - 7x + 9$ near   x = 4, first calculate the derivative $f'(x) = \underline{x - 7}$.   The value $f'(4) = \underline{-3}$ is the slope of the linearization at the point $(4, \underline{-11})$ on the graph of   f.   Thus, an equation of the linearization is $L(x) = -11 + \underline{(-3)}$ (x - 4),   or $L(x) = \underline{1 - 3x}$. $\quad -3x + 12$

$\quad\quad -20 + 9 = -11$
$\quad B \quad -28$

45.   From the result in Problem 44, an <u>estimate</u> to $\frac{1}{2}\left(\frac{1}{6}\right)^2 - \frac{7}{6} + 9$ is $\underline{-3\left(\frac{1}{6}\right) + 1}$. $= -\frac{1}{2} + 1 = \frac{1}{2}$

OBJECTIVE B :   Estimate the change   Δy   produced in a function   y = f(x)   when   x   changes by a small amount   dx.

46.   An estimate of   Δy   is given by the <u>differential</u> $dy = \underline{f'(x)dx}$.   Thus,   dy   denotes the change in the linearization of   f   that results from the change   dx   in   x.

47.   The equation $\frac{dy}{dx} = f'(x)$ says we may regard the derivative as a   <u>quotient</u> of differentials.

---

41. $x\frac{dy}{dx}$, $-\frac{y}{x}$, -1, (-1)(x - 1) or 1 - x.   42.   1, x - 1   43.   f(a) + f'(a)(x - a)

44.   x - 7, -3, -11, -3, -3x + 1   45. $-3\left(\frac{1}{6}\right) + 1 = \frac{1}{2}$

46.   f'(x) dx   47.   quotient

48.   Suppose we wish to estimate the change in $y = x^3$ when x
      changes by $dx = 0.1$ at $x = 2$. Now $\Delta y \approx dy = \dfrac{dy}{dx} dx$
      $= \underline{3x^2}\ dx$. When $x = 2$ and $dx = 0.1$,
      $dy = 3(\underline{\ 2\ })^2(\underline{\ .1\ }) = 1.2$. Therefore, since
      $f(x + dx) = y + \Delta y \approx y + dy$, $(2 + 0.1)^3 \approx 2^3 + \underline{1.2}$
      $= \underline{9.2}$. The actual value of $(2.1)^3$ is $\underline{9.261}$
      giving an error in our estimate of $\epsilon dx = \Delta y - dy = \underline{.061}$.
      The positive sign of ϵdx indicates that our estimate 9.2 is
      too small.

49.   To estimate the value of $\sqrt{16.56}$, let $y = \sqrt{x}$, $x = 16$, and
      $dx = 0.56$. Then $dy = (\underline{\ 1/2\sqrt{x}\ })dx$, so when $x = 16$ and
      $dx = 0.56$, $dy = (\underline{\ 1/8\ })(0.56) = .07$. Thus, $\sqrt{16.56} = \sqrt{16}$
      $+ .07 = \underline{\ 4.07\ }$. (The actual value of $\sqrt{16.56}$ is 4.0694
      correct to 5 decimal places, so our estimate is fairly
      accurate.)

2-5   THE CHAIN RULE.

OBJECTIVE :   If y is a differentiable function of x and x is a
              differentiable function of t, use the chain rule to
              calculate dy/dt.

50.   If y is a differentiable function of x and x is a
      differentiable function of t, then y is a differentiable
      function of $\underline{\ t\ }$, and $\dfrac{dy}{dt} = \underline{\dfrac{dy}{dx} \cdot \dfrac{dx}{dt}}$. This rule is known
      as the $\underline{\text{Chain}}$ rule for the derivative of a composite function.

51.   Consider the chain rule in functional form: let $y = g(x)$ and
      $x = f(t)$ be differentiable functions. Then the composite
      $y = h(t) = g(f(t))$ is a differentiable function of $\underline{\ t\ }$.
      When $t = t_0$, let $x = f(t_0) = x_0$. According to the chain
      rule the derivative of h evaluated at $t = t_0$ is given by
      $h'(t_0) = \underline{g'(f(t_0))}$. In this equation, $h'(t_0)$ corresponds
      to $(dy/dt)_{t_0}$, $g'(x_0)$ corresponds to $\underline{(dy/dx)_{x_0}}$, and $\underline{f'(t_0)}$
      corresponds to $(dy/dt)_{t_0}$. It is important that you observe
      that the derivatives in the chain rule equation $\dfrac{dy}{dt} = \dfrac{dy}{dx} \cdot \dfrac{dx}{dt}$
      are being evaluated at different points: dy/dt and dx/dt
      are evaluated at $\underline{\ t_0\ }$, whereas dy/dx is evaluated at
      $f(t_0) = \underline{\ x_0\ }$. Failure to understand this fact can lead to
      serious misuse of the chain rule equation.

---

48. $3x^2$, 2, 0.1, 1.2, 9.2, 9.261, 0.061   49. $\dfrac{1}{2\sqrt{x}}$, $\dfrac{1}{8}$, 16, 4.07

50. t, $\dfrac{dy}{dx} \cdot \dfrac{dx}{dt}$, chain   51. t, $g'(x_0) \cdot f'(t_0)$, $(dy/dx)_{x_0}$, $f'(t_0)$,

                                                    t_0, x_0

52. To find dy/dt if $y = x^2 - 2x + 3$ and $x = \sqrt{t}$, calculate $dy/dx = \underline{2x - 2}$ and then substitute $x = \sqrt{t}$ to obtain $(dy/dx)_{\sqrt{t}} = \underline{2\sqrt{t} - 2}$. According to the chain rule,

$$\frac{dy}{dt} = \frac{dy}{dx} \cdot \frac{dx}{dt} = (2\sqrt{t} - 2) \cdot \underline{\qquad} = \underline{1 - 1/\sqrt{t}} .$$

53. Suppose $y = z^{-2} + 3z^{-1}$ and $z = x^2 + 1$. Then $dy/dz = \underline{-2z^{-3} - 3z^{-2}}$ so that when $z = x^2 + 1$, $(dy/dz)_{x^2+1} = \underline{-2(x^2+1)^{-3} - 3(x^2+1)^{-2}}$. Applying the chain rule,

$$\frac{dy}{dx} = \left[-2\left(x^2 + 1\right)^{-3} - 3\left(x^2 + 1\right)^{-2} \right] \cdot \underline{2x}$$

$$= \frac{-2x}{\left(x^2 + 1\right)^2} \left[\frac{2}{x^2 + 1} + \underline{3} \right] = \frac{-2x(3x^2 + 5)}{\left(x^2 + 1\right)^3} .$$

54. Suppose $y = (2x^3 - x^2 + 3x - 1)^4$. Then

$$\frac{dy}{dx} = \underline{4(2x^3 - x^2 + 3x - 1)^3(6x^2 - 2x + 3)}$$

2-6 A BRIEF REVIEW OF TRIGONOMETRY.

Problems 55-61 give the most important trigonometric formulas to remember.

55. $\sin (A + B) = \underline{\sin A \cos A + \sin B \cos B}$ 56. $\cos (A + B) = \underline{\cos A \cos B - \sin A \sin B}$

57. $\sin (-A) = \underline{-\sin A}$. 58. $\cos (-A) = \underline{\cos A}$.

59. Referring to the triangle ABC at the right, the <u>law</u> <u>of</u> <u>cosines</u> states that

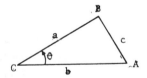

$c^2 = \underline{a^2 + b^2 - 2ab \cos \theta}$.

60. $\sin^2\theta + \cos^2\theta = \underline{1}$.

61. Dividing both sides of the equation in Problem 60 by $\cos^2\theta$ gives $\tan^2\theta + 1 = \underline{\sec^2\theta}$.

---

52. $2x - 2$, $2\sqrt{t} - 2$, $\frac{1}{2\sqrt{t}}$, $1 - \frac{1}{\sqrt{t}}$

53. $-2z^{-3} - 3z^{-2}$, $-2\left(x^2 + 1\right)^{-3} - 3\left(x^2 + 1\right)^{-2}$, $2x$, 3, $-2x(3x^2 + 5)$

54. $4(2x^3 - x^2 + 3x - 1)^3(6x^2 - 2x + 3)$ 55. $\sin A \cos B + \cos A \sin B$

56. $\cos A \cos B - \sin A \sin B$ 57. $- \sin A$

58. $\cos A$ 59. $a^2 + b^2 - 2ab \cos \theta$ 60. 1

61. $\sec^2\theta$

OBJECTIVE :   Given a general sine function of the form

$$f(x) = A \sin\left[\frac{2\pi}{B}(x - C)\right] + D$$

identify the amplitude, period, horizontal shift, and vertical shift. Sketch the graph of the function.

62.   Consider the function $f(x) = \frac{1}{2} \sin(3x - 2) + \frac{1}{2}$. Here the amplitude A = ___ 1/2 ___ . To find the period, we need to write $(3x - 2)$ in the form $\frac{2\pi}{B}(x - C)$, and we proceed as follows:

$$3x - 2 = 3(\underline{x - \tfrac{2}{3}}) = \underline{\tfrac{1}{\tfrac{1}{3}}}\left(x - \tfrac{2}{3}\right) = \underline{\tfrac{2\pi}{\tfrac{2\pi}{3}}}\left(x - \tfrac{2}{3}\right).$$

Therefore, the period B = ___ $2\pi/3$ ___, the horizontal shift C = ___ $2/3$ ___, and the vertical shift D = ___ $1/2$ ___. Observe that when $x = C$, the sine of $\frac{2\pi}{B}(x - C)$ is zero. Thus, when C is ___ positive ___ the graph of f is shifted to the right, and when C is negative, it is shifted to the ___ left ___ . Sketch one cycle of the graph of f in the coordinate system below.

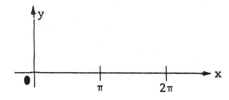

2-7   DERIVATIVES OF TRIGONOMETRIC FUNCTIONS.

OBJECTIVE :   Calculate the derivatives of functions involving the trigonometric functions, making use of appropriate rules of differentiation and the derivatives of the sine and cosine functions.

63. $\frac{d}{dx} \sin x$ = ___ cos x ___ .

64. $\frac{d}{dx} \cos x$ = ___ -sin x ___ .

---

62. $\frac{1}{2}$, $x - \frac{2}{3}$, $\frac{1}{3}$, $\frac{2\pi}{3}$, $\frac{2\pi}{3}$, $\frac{2}{3}$, $\frac{1}{2}$,

  positive, left

63.   cos x

64.   - sin x

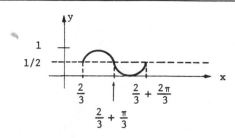

65. $\frac{d}{dx} \tan x = \frac{d}{dx} \frac{\sin x}{\cos x} = \frac{\cos x \cdot \frac{d}{dx}(\sin x) - \sin x \frac{d}{dx}(\cos x)}{\cos^2 x}$

$= \frac{\cos x \,(+\cos x) - \sin x \,(-\sin x)}{\cos^2 x}$

$= \frac{\cos^2 x + \sin^2 x}{\cos^2 x} = \underline{\sec^2 x}$.

66. $\frac{d}{dx} \sec x = \frac{d}{dx} \frac{1}{\cos} = (-)\frac{1}{\cos^2 x} \cdot \frac{d}{dx}(\cos x)$

$= \frac{+\sin x}{\cos^2 x} = \underline{\tan x \sec x}$.

Remark: It will be to your advantage in later work to MEMORIZE the derivative formulas derived in Problems 65 and 66.

67. $\frac{d}{dx} x \tan^2 x = x \frac{d}{dx}(\tan^2 x) + \frac{dx}{dx} \tan^2 x = x \cdot 2 \tan \frac{d}{dx}(\tan x) + \tan^2 x$

$= 2x \,\underline{\tan x \sec^2 x} + \tan^2 x$

$= \underline{\tan x (2x \sec^2 x + \tan x)}$

68. $\frac{d}{dx} \sqrt{\frac{1 - \cos x}{1 + \cos x}} = \frac{1}{2}\left(\frac{1-\cos x}{1+\cos x}\right)^{-1/2} \frac{d}{dx}\left(\frac{1 - \cos x}{1 + \cos x}\right)$

$= \frac{1}{2}\left(\frac{1 - \cos x}{1 + \cos x}\right)^{-1/2}\left[\frac{(1 + \cos x)(\sin x) - (1 - \cos x)(-\sin x)}{(1 + \cos x)^2}\right]$

$= \frac{(1 - \cos x)^{-1/2} \sin x}{(1+\cos x)^{3/2}}$.

69. The linearization of the function $f(x) = \sin x$ at $x = \frac{\pi}{2}$ is obtained using the equation

$$L(x) = f(a) + f'(a)(x - a)$$

when $a = \underline{\pi/2}$. Now $f'(x) = \underline{\cos x}$ so that $f'(\frac{\pi}{2}) = \underline{\cos \pi/2} = 0$ Thus, since $f(\frac{\pi}{2}) = \underline{1}$ the linearization of f at $\frac{\pi}{2}$ is given by

$$L(x) = \underline{1 + 0\left(x - \pi/2\right)} = 1 .$$

Thus, near $x = \frac{\pi}{2}$, $\sin x \approx \underline{1}$.

---

65. $\cos x$, $\sin x$, $\cos x$, $-\sin x$, $\sin^2 x$, $\sec^2 x$ 66. $\cos x$, $\cos' x$, $\sin x$, $\sec x \tan x$

67. $\tan^2 x$, $\tan x$, $\tan x \sec^2 x$, $\tan x(2x \sec^2 x + \tan x)$

68. $\left(\frac{1 - \cos x}{1 + \cos x}\right)^{-1/2}$, $\sin x$, $-\sin x$, $(1 + \cos x)^{3/2}$

69. $\pi/2$, $\cos x$, 0, 1, $1 + 0(x - \frac{\pi}{2})$, 1

2-8   PARAMETRIC EQUATIONS.

OBJECTIVE : Given parametric equations $x = f(t)$ and $y = g(t)$,
eliminate the parameter t to find an equation in the
form $y = F(x)$. Find dy/dx in terms of dy/dt and
dx/dt. Find d^2y/dx^2 in terms of t.

70. The equations $x = f(t)$ and $y = g(t)$, which express x and
y in terms of t, are called *parametric* equations. The
variable t is called a *parameter* . From the chain rule,
the derivative dy/dx is given by $\dfrac{dy}{dx} = \dfrac{dy/dt}{dx/dt}$.

71. Consider the curve given by the parametric equations $x = t^2$
and $y = t^2 - 2t$. First complete the following table of values
for the coordinates of the path:

t	-2	-1	0	1	2	3
x	4	1	0	1	4	9
y	8	3	0	-1	0	3

Plot a graph of the curve in the
coordinate system at the right.

Differentiating, $\dfrac{dx}{dt} = 2t$
and $\dfrac{dy}{dt} = 2t - 2$. Thus,
$\dfrac{dy}{dt} = $ $1 - \frac{1}{t}$ Note that dy/dx
fails to exist when $t = 0$. This is evidenced in the
graph by observing that the curve is tangent to the y-axis when
t = 0 at the point (0,0). When t = 1, dy/dx = 0 .
Thus the curve has a *horizontal* tangent at (1,-1). Note
that the x-coordinate is non-negative for all values of t.
Also, $\dfrac{d^2y}{dx^2} = \dfrac{dy'/dt}{dx/dt} = \dfrac{1/t^2}{2t} = \dfrac{1}{2t^3}$.

---

70. parametric, parameter, $\dfrac{dy/dt}{dx/dt}$

71.

t	-2	-1	0	1	2	3
x	4	1	0	1	4	9
y	8	3	0	-1	0	3

$\dfrac{dx}{dt} = 2t$, $\dfrac{dy}{dt} = 2(t-1)$, $\dfrac{dy}{dx} = 1 - \dfrac{1}{t}$,

0, 0, horizontal, dy'/dt, $1/t^2$, $\dfrac{1}{2t^3}$

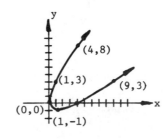

72. To eliminate the parameter t in Problem 71, observe that $t = \sqrt{x}$ if $t \geq 0$, and $t = \underline{-\sqrt{x}}$ if $t < 0$. Thus

$$y = \begin{cases} \underline{x - 2\sqrt{x}}, & \text{if } t \geq 0 \\ \underline{x + 2\sqrt{x}}, & \text{if } t < 0 \end{cases}$$

2-9 NEWTON'S METHOD FOR APPROXIMATING SOLUTIONS OF EQUATIONS.

OBJECTIVE: Use the Newton-Raphson method to estimate the root of an equation $f(x) = 0$ within specified $a \leq x \leq b$.

73. In using the Newton-Raphson method, to go from the nth approximation x_n of the root to the next approximation x_{n+1}, use the formula $x_{n+1} = \underline{x_n - \dfrac{f(x_n)}{f'(x_n)}}$. This formula fails if the derivative $f'(x_n)$ equals $\underline{0}$.

74. Suppose it is required to find a real root to the equation $f(x) = x^3 + x - 1$. Since $f(0) = \underline{-1}$ and $f(1) = \underline{1}$ differ in sign, an unknown root lies somewhere in the interval $0 < x < 1$. As a first guess, choose $x_1 = 0.5$. To apply the Newton-Raphson formula, we calculate the derivative $f'(x) = \underline{3x^2 + 1}$. Then,

$$x_2 = x_1 - \frac{x_1^3 + x_1 - 1}{3x_1^2 + 1} = \frac{1}{2} - \frac{(1/2)^3 + (1/2) - 1}{3(1/2)^2 + 1} = \frac{1}{2} + \frac{3}{14}$$

$$= \frac{5}{7} \approx 0.71429.$$

$$x_3 = x_2 - \frac{x_2^3 + x_2 - 1}{3x_2^2 + 1} = \frac{5}{7} - \frac{(5/7)^3 + (5/7) - 1}{3(5/7)^2 + 1}$$

$$= \frac{5}{7} - \frac{27}{7 \cdot 124} = \frac{593}{7 \cdot 124} \approx 0.68318.$$

With the aid of a hand-held programmable calculator, we have computed the following iterations in the same way: $x_4 = 0.68233$ and $x_5 = 0.68233$. Thus, a root to $f(x) = x^3 + x - 1$ is $r = 0.68233$ correct to 5 decimal places. The method is easy, but the arithmetic can be cumbersome without the aid of a calculator.

---

72. $-\sqrt{x}$, $\begin{cases} x - 2\sqrt{x} \\ x + 2\sqrt{x} \end{cases}$

73. $x_n - \dfrac{f(x_n)}{f'(x_n)}$, 0

74. -1, 1, $3x^2 + 1$, $3(1/2)^2 + 1$, 3, 5, $x_2^3 + x_2 - 1$, $(5/7)^3 + (5/7) - 1$, 27, 593

2-10 DERIVATIVE FORMULAS IN DIFFERENTIAL NOTATION.

OBJECTIVE A : Given $y = f(x)$, find the differential dy.

75.   If $y = x^2 + \sin 3x$, then $\dfrac{dy}{dx} =$ _____ . Thus,

$dy =$ _____ .

OBJECTIVE B : Given $x = f(t)$ and $y = g(t)$, express dx and dy
in terms of t and dt.

76.   Let $x = t^2 - 1$ and $y = 1/t$. Then $\dfrac{dx}{dt} =$ _____ and

$\dfrac{dy}{dt} =$ _____ . It follows that $dx = 2t\,dt$ and $dy = -dt/t^2$.

---

75. $2x + 3\cos 3x$, $(2x + 3\cos 3x)\,dx$

76. $2t$, $-1/t^2$

CHAPTER 2   SELF-TEST

1. Find $\dfrac{dy}{dx}$

    (a) $y = (2x^3 - x^2 + 7x + 3)^6$

    (b) $y = (x^2 - 9)(3x^5 + 7x)$

    (c) $y = \dfrac{3}{x^4 + 1}$

    (d) $y = \dfrac{x^2 - 1}{3x + 1}$

2. Find $\dfrac{d^2y}{dx^2}$

    (a) $y = \frac{1}{3}x^3 + \frac{1}{2}x^2 - 6x + 8$

    (b) $y = (2x^3 - 11)(x^2 - 3)$

3. A particle moves along a horizontal line (positive to the right) according to the law

    $$s = t^3 - 6t^2 + 2.$$

    During which intervals of time is the particle moving to the right and during which is it moving to the left? What is the acceleration and the velocity when $t = 2.3$?

4. Find an equation of the tangent line to the graph of $y = \sqrt{1 - x^2}$ when $x = 1/2$.

5. Find $\dfrac{dy}{dx}$

    (a) $y = x^{4/3} - 5x^{4/5}$

    (b) $y = \sqrt{x + \frac{1}{x}}$

    (c) $y = \dfrac{1}{2x - \sqrt{x^2 - 1}}$

    (d) $y = x \cos(5x - 2)$

6. Find the linearization of $f(x) = \sqrt{x + \frac{1}{x}}$ at $x = 4$.

7. Find $\dfrac{dy}{dx}$ and $\dfrac{d^2y}{dx^2}$ when $x + y^2 = xy$.

8. Find an equation of the tangent line to the curve $x^3 + 3xy^3 + xy^2 = xy$ at the point $(1,-1)$.

9. Use the linearization $L(x)$ to estimate the value of $\sin 29°$.

10. Beginning with the estimate $x_1 = \frac{\pi}{2}$, apply Newton's method once to calculate a positive solution to the equation $\sin x = \frac{2}{3}x$.

11. Consider the curve given by the parametric equations $x = 3t - 1$ and $y = t^2 - t$. Eliminate the parameter t to find an equation of the form $y = F(x)$ and find dy/dx in terms of dy/dt and dx/dt. For what value of t is $dy/dx = 0$? Sketch the graph of the curve over the interval $-2 \leq t \leq 3$.

12. Identify the amplitude, period, horizontal shift, and vertical shift of the function $f(x) = 2 \sin(3x + 1)$. Sketch one cycle of the graph of the function.

13.   Find $\dfrac{dy}{dx}$

    (a) $y = \cot 3x$ (b) $y = 3 \sin^2 5x - \sec x$

14.   Given the parametric equations $x = t^2 - 1$ and $y = t + 1$

    (a)   Express dx and dy in terms of t and dt,

    (b)   Find d^2y/dx^2 in terms of t,

    (c)   Find an equation of the tangent line to the curve at the point for which   t = 1.

SOLUTIONS TO CHAPTER 2 SELF-TEST

1.   (a) $\dfrac{dy}{dx} = 6(2x^3 - x^2 + 7x + 3)^5 \, (6x^2 - 2x + 7)$

    (b) $\dfrac{dy}{dx} = 2x(3x^5 + 7x) + (x^2 - 9)(15x^4 + 7)$

                 $= 21x^6 - 135x^4 + 21x^2 - 63$

    (c) $\dfrac{dy}{dx} = \dfrac{-3(4x^3)}{(x^4 + 1)^2}$

    (d) $\dfrac{dy}{dx} = \dfrac{(3x + 1)(2x) - (x^2 - 1)(3)}{(3x + 1)^2} = \dfrac{3x^2 + 2x + 3}{(3x + 1)^2}$

2.   (a) $\dfrac{dy}{dx} = x^2 + x - 6, \quad \dfrac{d^2y}{dx^2} = 2x + 1$

    (b) $\dfrac{dy}{dx} = 6x^2(x^2 - 3) + (2x^3 - 11)(2x) = 10x^4 - 18x^2 - 22x,$

       $\dfrac{d^2y}{dx^2} = 40x^3 - 36x - 22$

3. $\dfrac{ds}{dt} = 3t^2 - 12t = 3t(t - 4); \quad \dfrac{d^2s}{dt^2} = 6t - 12$

    The particle is moving to the right when $\dfrac{ds}{dt} > 0$ so t > 4 or t < 0;   it is moving to the left when 0 < t < 4 and $\dfrac{ds}{dt} < 0$. At   t = 2.3,

$$\left.\dfrac{ds}{dt}\right|_{2.3} = 3(2.3)^2 - 12(2.3) = -11.73, \quad \text{velocity}$$

$$\left.\dfrac{d^2s}{dt^2}\right|_{2.3} = 6(2.3) - 12 = 1.8, \quad \text{acceleration}$$

4. $y' = \frac{1}{2}(1 - x^2)^{-1/2}(-2x) = -x(1 - x^2)^{-1/2}, \quad y(\frac{1}{2}) = \frac{\sqrt{3}}{2}$ and

    $y'(\frac{1}{2}) = -\dfrac{1}{\sqrt{3}}$ so that $\left(y - \frac{\sqrt{3}}{2}\right) = -\dfrac{1}{\sqrt{3}}\left(x - \frac{1}{2}\right)$ or

    $\sqrt{3}\,y + x - 2 = 0$ is an equation of the tangent line.

5. (a) $\dfrac{dy}{dx} = \dfrac{4}{3}x^{1/3} - 4x^{-1/5}$

(b) $\dfrac{dy}{dx} = \dfrac{1}{2}\left(x + \dfrac{1}{x}\right)^{-1/2} \cdot \dfrac{d}{dx}\left(x + \dfrac{1}{x}\right) = \dfrac{1}{2}\left(x + \dfrac{1}{x}\right)^{-1/2}\left(1 - \dfrac{1}{x^2}\right)$

(c) $\dfrac{dy}{dx} = -\left(2x - \sqrt{x^2 - 1}\right)^{-2}\left[2 - \dfrac{1}{2}(x^2 - 1)^{-1/2} \cdot 2x\right]$

$= -\left(2x - \sqrt{x^2 - 1}\right)^{-2}\left(2 - \dfrac{x}{\sqrt{x^2 - 1}}\right)$

(d) $\dfrac{dy}{dx} = \cos(5x - 2) + x[-\sin(5x - 2)\cdot 5]$

$= \cos(5x - 2) - 5x\,\sin(5x - 2)$

6. From Problem 5(b), $f'(4) = \dfrac{1}{2}\left(4 + \dfrac{1}{4}\right)^{-1/2}\left(1 - \dfrac{1}{16}\right) = \dfrac{15}{16\sqrt{17}} \approx 0.227$

and $f(4) = \sqrt{\dfrac{17}{4}} \approx 2.062$. Thus, $L(x) = \dfrac{\sqrt{17}}{2} + \dfrac{15}{16\sqrt{17}}(x - 4)$

$\approx 2.062 + 0.227(x - 4)$.

7. Differentiating implicitly, $1 + 2yy' = y + xy'$ or $y' = \dfrac{y - 1}{2y - x}$

$\dfrac{d^2y}{dx^2} = \dfrac{(2y - x)(y') - (y - 1)(2y' - 1)}{(2y - x)^2} = \dfrac{(2 - x)y' + (y - 1)}{(2y - x)^2}$

$= \dfrac{(2 - x)\left(\dfrac{y - 1}{2y - x}\right) + (y - 1)}{(2y - x)^2} = \dfrac{2(y - 1)(y - x + 1)}{(2y - x)^3}$.

8. Since $1^3 + 3(1)(-1)^3 + 1(-1)^2 = 1(-1)$ is true, the point $(1,-1)$
is on the curve. Differentiating implicitly,
$3x^2 + 3y^3 + 9xy^2y' + y^2 + 2xyy' = y + xy'$, so evaluation at
$(1,-1)$ yields $3 - 3 + 9y' + 1 - 2y' = -1 + y'$, or $y' = -\dfrac{1}{3}$.
Thus an equation of the tangent line is given by

$$(y + 1) = -\dfrac{1}{3}(x - 1) \text{ or } x + 3y = -2 .$$

9. The calculation must be done when $y = \sin x$ for x measured in
<u>radians</u>. Thus,

$$\sin 29° \approx \sin \dfrac{\pi}{6} + dy,$$

where $dy = \dfrac{dy}{dx}\,dx$ when $x = \dfrac{\pi}{6}$ and $dx = -\dfrac{\pi}{180}$ radians.

Now, $\dfrac{dy}{dx}\Big|_{\pi/6} = \cos \dfrac{\pi}{6} = \dfrac{\sqrt{3}}{2}$ so that

$$\sin 29° \approx \dfrac{1}{2} + \left(\dfrac{\sqrt{3}}{2}\right)\left(-\dfrac{\pi}{180}\right) \approx .48489.$$

10. Let $f(x) = \sin x - \dfrac{2}{3}x = 0$, $f'(x) = \cos x - \dfrac{2}{3}$. By Newton's

method, $x_2 = x_1 - \dfrac{\sin x_1 - (2/3)x_1}{\cos x_1 - (2/3)} = \dfrac{\pi}{2} - \dfrac{1 - \pi/3}{0 - 2/3}$

$= \dfrac{\pi}{2} + \dfrac{3}{2}\left(1 - \dfrac{\pi}{3}\right) = 1.5$.

11. $\frac{dx}{dt} = 3$ and $\frac{dy}{dt} = 2t - 1$,   so that $\frac{dy}{dx} = \frac{dy/dt}{dx/dt} = \frac{1}{3}(2t - 1)$.   When $t = \frac{1}{2}$, $\frac{dy}{dx} = 0$ which occurs at the point $(x,y) = (\frac{1}{2}, -\frac{1}{4})$.   To eliminate the parameter   t,   use the parametric expression for   x, $t = \frac{1}{3}(x + 1)$,   and substitute into the parametric expression for   y   giving

$y = \frac{1}{9}(x + 1)^2 - \frac{1}{3}(x + 1)$

  $= \frac{1}{9}(x^2 - x - 2)$.

A table of coordinate values is as follows:

t	-2	-1	0	1/2	1	2	3
x	-7	-4	-1	1/2	2	5	8
y	6	2	0	-1/4	0	2	6

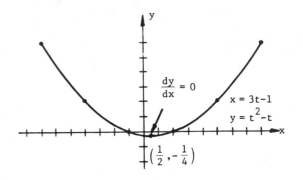

12. $2 \sin(3x + 1) = 2 \sin\left[\frac{2\pi}{2\pi/3}\left(x + \frac{1}{3}\right)\right]$

   amplitude = 2;   period = $\frac{2\pi}{3}$; horizontal shift = $-\frac{1}{3}$ (a shift to the left); vertical shift = 0.

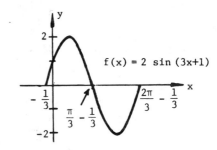

13. (a) $\frac{dy}{dx} = \frac{d}{dx}\left(\frac{\cos 3x}{\sin 3x}\right) = \frac{\sin 3x(-3 \sin 3x) - (\cos 3x)(3 \cos 3x)}{\sin^2 3x}$

      $= \frac{-3 \sin^2 3x - 3 \cos^2 3x}{\sin^2 3x} = -3 \csc^2 3x$.

   (b) $\frac{dy}{dx} = 30 \sin 5x \cos 5x - \sec x \tan x$

14. (a)   dx/dt = 2t   and   dy/dt = 1,   so that   dx = 2t dt   and dy = dt.

   (b) $\frac{dy}{dx} = \frac{1}{2t}$ so that $\frac{d^2y}{dx^2} = \frac{dy'/dt}{dx/dt} = \frac{-1/2t^2}{2t} = -\frac{1}{4t^3}$

   (c)   When   t = 1,   x = 0,   y = 2,   and $\frac{dy}{dx} = \frac{1}{2}$;   thus $y - 2 = \frac{1}{2}(x - 0)$ or $2y - x = 4$ is an equation of the tangent line.

CHAPTER 3 APPLICATIONS OF DERIVATIVES

3-1 SKETCHING CURVES WITH THE FIRST DERIVATIVE.

OBJECTIVE A: Use the sign of the first derivative dy/dx to determine the values of x where the graph of y versus x is increasing and where it is decreasing.

1. Let $y = f(x)$ be a differentiable function of x. When dy/dx has a _positive_ value, the graph of y versus x is rising (to the right). In this case it is also said that the function f is _increasing_ at that point.

2. When dy/dx < 0, the graph of y versus x is ~~_falling_~~ and the function f is _decreasing_ at that point.

3. Let $y = \frac{1}{3}x^3 - x^2 + 2$. Then $y' = \underline{x^2 - 2x} = x(\underline{x - 2})$. The derivative dy/dx is zero when $x = \underline{2}$ or $x = \underline{0}$. Thus, the curve y is increasing when x < 0, it is decreasing when x satisfies $\underline{0 < x < 2}$, and it is increasing again when $x > \underline{2}$. We construct a table of some values for the curve (complete the table):

$-\frac{1}{3} - \frac{3}{3} + \frac{6}{3}$

$-\frac{1}{3}, \frac{3}{3} = \frac{2}{3}$

x	-2	-1	0	1	2	3	4
y	14/3	2/3	2	4/3	2/3	2	22/3

Sketch the graph in the coordinate system at the right.

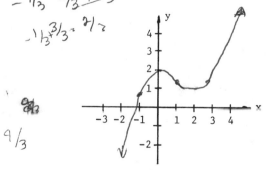

$\frac{1}{3} - \frac{3}{3} + \frac{4}{3}$

$-\frac{2}{3} + \frac{4}{3} = \frac{9}{3}$

27

---

1. positive, increasing

2. falling, decreasing

3. $x^2 - 2x$, x - 2, 0, 2, 0 < x < 2, 2

x	-2	-1	0	1	2	3	4
y	-14/3	2/3	2	4/3	2/3	2	22/3

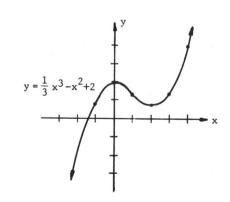

$y = \frac{1}{3}x^3 - x^2 + 2$

4. The following graph describes a differentiable function
   $y = f(x)$.

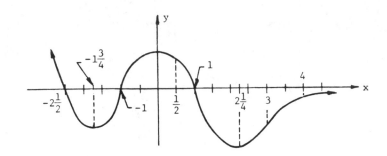

The function f is positive when x belongs to the interval
$(-\infty, -2\frac{1}{2})$ or the interval $(-1, 1)$. There are no points of
discontinuity of f. When $x = 0$ or $x = -1\frac{3}{4}$ or
$x = 2\frac{1}{4}$, the derivative f' changes sign. The
derivative is positive when x belongs to the interval
$(-\frac{3}{4}, 0)$ or the interval $(2\frac{1}{4}, \infty)$; the derivative is
negative when x belongs to $(-\infty, -5/2)$ or $0, 2\frac{1}{4}$.
Thus, the curve is rising for $-7/4 < x < 0$ and $x > 9/4$,
and it is falling for $x < -7/4$ and $0 \le x \le 2\frac{1}{4}$

OBJECTIVE B : Given a function $y = f(x)$, find the intervals of
values of x for which the curve is increasing or
decreasing. Find any local maximum and local minimum
values the function has when $y' = 0$. Sketch the curve.

5. Consider the function $y = 2 + 3x - x^3$. Then $y' = 3 - 3x^2$
   so that $y' = 0$ when $x = 1$ and $x = -1$. For
   $-1 < x < 1$ the curve is increasing for $x < -1$ and $x > 1$
   the curve is decreasing. At $x = -1$ the curve has a local
   minimum; at $x = 1$ the curve has a local maximum. We

---

4. $(-\infty, -5/2)$, $(-1, 1)$, $-7/4$, 0, $9/4$, $(-7/4, 0)$,

   $(9/4, \infty)$, negative, $(0, 9/4)$, rising, $x < -7/4$,

   $0 < x < 9/4$

5. $3 - 3x^2$, -1, 1, increasing, 1, minimum, 1

x	-2	-1	0	1	2	3
y	4	0	2	4	0	-16

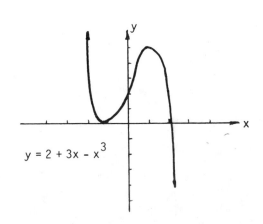

$y = 2 + 3x - x^3$

construct a table of some values for the curve (complete the table):

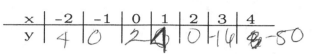

x	-2	-1	0	1	2	3	4
y	4	0	2	4	0	-16	-50

Sketch the graph in the coordinate system at the right.

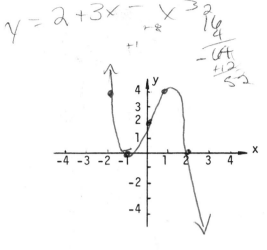

$y = 2 + 3x - x^3$

3-2 CONCAVITY AND POINTS OF INFLECTION.

OBJECTIVE A : Relate the concavity of a function $y = f(x)$ to the second derivative d^2y/dx^2.

6. If the second derivative d^2y/dx^2 is positive, the y-curve is concave ___up___ at that point; if d^2y/dx^2 is ___negative___, the curve is concave downward at that point.

7. When a curve is concave upward at a point, locally the curve lies _____ the tangent line; when it is concave _____, locally the curve lies below the tangent line.

8. A point where the curve changes concavity is called a _____ _____, and is characterized by a change in sign of _____.

9. A point of inflection occurs where d^2y/dx^2 is _____ or _____.

10. Does the condition $d^2y/dx^2 = 0$ guarantee a point of inflection? _____

---

6. upward, negative

7. above, downward

8. point of inflection, d^2y/dx^2

9. zero, fails to exist

10. No, the function $y = x^4$ affords a counterexample at $x = 0$.

OBJECTIVE B : Given a function $y = f(x)$, find the intervals of values of x for which the curve is rising, falling, concave upward, and concave downward. Sketch the curve, showing the points of inflection and the points where the function has local maximum and local minimum values.

11. Consider the function $f(x) = x^4 - 4x^3 + 10$. We follow a five-step procedure in order to sketch the graph of $y = f(x)$.

(a) $\dfrac{dy}{dx} =$ _____ and $\dfrac{d^2y}{dx^2} =$ _____ .

(b) In factored terms, $dy/dx = 4x^2(x - 3)$ so that the curve is decreasing when x belongs to the interval _____ and increasing when $x >$ _____. The slope of the curve is zero when $x =$ _____ or $x =$ _____.

(c) In factored form, $d^2y/dx^2 =$ _____. Thus d^2y/dx^2 is negative when x belongs to the interval _____ and consequently the curve is concave _____ there. The second derivative is positive when x satisfies _____ or _____, and the curve is concave _____. Therefore, the second derivative changes sign when $x =$ _____ or $x =$ _____ so that these are points of inflection of f.

(d) Complete the following table.

x	y	y'	y''	Conclusions
-2	58	-	+	falling; concave up
-1	15	-	+	falling; concave up
0				
1				
2				
3				
4	20	+	+	rising; concave up

(e) Sketch a smooth curve of $y = f(x)$ in the given coordinate system to the right.

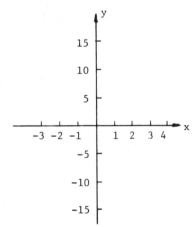

---

11. (a) $4x^3 - 12x^2$, $12x^2 - 24x$

(b) $(-\infty, 3)$, 3, 0, 3

(c) $12x(x - 2)$, $(0,2)$, downward, $x < 0$, $x > 2$, upward, 0, 2

(d)

x	y	y'	y''	Conclusions
0	10	0	0	point of inflection
1	7	-	-	falling; concave down
2	-6	-	0	point of inflection
3	-17	0	+	"Holds water"; min.

(e)

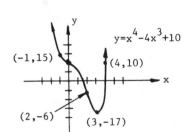

12. Sketch the graph of $y = x^{-2} + 2x$.
     <u>Solution</u>. We follow the five steps as in Problem 11.

(a) $y' =$ _____ and $y'' =$ _____ .

(b)   In fractional form, $y' = \dfrac{\rule{2cm}{0.4pt}}{x^3}$ so that y' is zero
      when $x =$ _____ . The curve is decreasing when x
      belongs to the interval _____ ; it is increasing for x
      satisfying _____ and _____ .

(c) d^2y/dx^2 is always _____ so the curve is everywhere
      concave _____ . Therefore there are no points of
      inflection.

(d)   The curve is discontinuous at
      $x =$ _____ . For large
      values of $|x|$, the curve is
      approximately $y \approx$ _____ .
      When x is small, the curve
      is approximately $y \approx$ _____ .

      Complete the following table.

x	y	y'	y''	Conclusions
-2	-15/4	+	+	rising; concave up
-1	-1			
-1/2	3			
1/2	5			
1				
2	17/4	+	+	rising; concave up

(e)   Sketch the graph at the right.

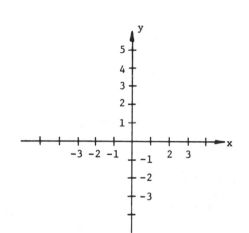

---

12. (a) $-2x^{-3} + 2,\; 6x^{-4}$ (e)

     (b) $2(x^3 - 1),\; 1,\; (0,1),\; x < 0,\; x > 1$

     (c) positive, upward

     (d) $0,\; 2x,\; 1/x^2$

x	y	y'	y''	Conclusions
-1	-1	+	+	rising; concave up
-1/2	3	+	+	rising, concave up
1/2	5	-	+	falling; concave up
1	3	0	+	min.; concave up

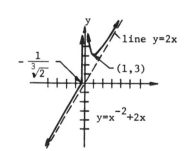

3-3   ASYMPTOTES AND SYMMETRY.

Questions 13-17 pertain to the graph of the equation $F(x,y) = 0$.

13.   The graph is symmetric about the _____ if $F(x,y) = F(y,x)$.

14.   The graph is symmetric about the _____ if $F(x,y) = F(-x,-y)$.

15.   The graph is symmetric about the _____ if $F(x,y) = F(-x,y)$.

16.   The graph is symmetric about the _____ if $F(x,y) = F(-y,-x)$.

17.   The graph is symmetric about the _____ if $F(x,y) = F(x,-y)$.

18.   A line $y = b$ is a _____ asymptote of the graph of a function $y = f(x)$ if either $\lim\limits_{x \to \infty} f(x) = b$ or
_____.

[OBJECTIVE]:   Analyze a given function $y = f(x)$ to investigate the following properties of the curve: (a) symmetry   (b) intercepts   (c) asymptotes   (d) slope at the intercepts   (e) rise and fall   (f) concavity, and   (g) dominant terms.   Using the information you have discovered, sketch the curve.

19.   Let $y = \dfrac{3x^2 - 1}{x^3}$.   Since $-y = \dfrac{3(-x)^2 - 1}{(-x)^3}$,   the curve is

symmetric about the _____.   Moreover, y is undefined when $x =$ _____,   but the graph is defined for all other real values of x.   Now, $y = 0$ when $3x^2 - 1 = 0$.   Thus, the x-intercepts occur at _____.   Since $x \neq 0$,   there are no y-intercepts.   Now

$$\lim_{x \to 0^-} \frac{3x^2 - 1}{x^3} = \lim_{x \to 0^-} \left(\frac{3}{x} - \underline{\quad}\right) = \underline{\quad\quad}, \quad \text{and}$$

$$\lim_{x \to 0^+} \frac{3x^2 - 1}{x^3} = \underline{\quad\quad} .$$

Therefore, the line _____ is a _____ asymptote.   Also,

---

13.   line $y = x$ 14.   origin

15.   y-axis   16.   line $y = -x$

17.   x-axis   18.   horizontal, $\lim\limits_{x \to -\infty} f(x) = b$

19.   origin, 0, $x = \pm \sqrt{3}$, $1/x^3$, ∞, $-\infty$, $x = 0$,
vertical, 0, $y = 0$, horizontal, $3x^2 - 1$, x^4,
± 1, positive, $x > 1$, $x < -1$.

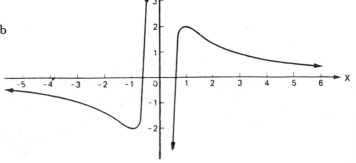

$$\lim_{x \to \pm\infty} \frac{3x^2 - 1}{x^3} = \underline{\hspace{3cm}}$$

so the line _____ is a _____ asymptote.

Finally, $y' = \dfrac{x^3(6x) - (\underline{\hspace{1.5cm}})3x^2}{x^6} = \dfrac{3(1 - x^2)}{\underline{\hspace{1cm}}}$. Thus, $y' = 0$

when $x = \underline{\hspace{2cm}}$. The derivative is _____ for $0 < x < 1$ and negative for _____ and for _____. A sketch of the graph is shown below.

20. Sketch the graph of $y = x^{1/3} + x^{2/3}$.

Solution. We follow the five steps as before.

(a) $y' = \underline{\hspace{3cm}}$ and $y'' = \underline{\hspace{2.5cm}}$.

(b) In factored form, $y' = \frac{1}{3} x^{-2/3} (\underline{\hspace{1.5cm}})$. Therefore, $y' = 0$ when $x = \underline{\hspace{2cm}}$. We observe that y' fails to exist when $x = \underline{\hspace{2cm}}$. However, $dx/dy = \underline{\hspace{1.5cm}}$ which equals zero when $x = 0$. Thus, the tangent to the curve at $(0,0)$ is _____.

(c) In factored form, $y'' = -\frac{2}{9} x^{-5/3}(\underline{\hspace{1cm}})$. Therefore, y'' is positive for _____ and negative whenever x belongs to the intervals _____ or _____. Hence the points $x = \underline{\hspace{1.5cm}}$ and $x = \underline{\hspace{1.5cm}}$ are inflection points.

(d) Complete the table:

x	y	y'	y''	Conclusions
-2		-	-	falling, concave down
-1				
-1/8				
0				
1	2	+	-	rising; concave down

(c) Sketch the graph. Observe that y'' fails to exist at the vertical tangent when $x = 0$. Note, however, that the curve is everywhere continuous.

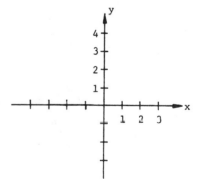

---

20. (a) $\frac{1}{3} x^{-2/3} + \frac{2}{3} x^{-1/3}$, $-\frac{2}{9} x^{-5/3} - \frac{2}{9} x^{-4/3}$

(b) $1 + 2x^{1/3}$, $-\frac{1}{8}$, 0, $\dfrac{3x^{2/3}}{1 + 2x^{1/3}}$, vertical

(c) $1 + x^{1/3}$, $-1 < x < 0$
$x < -1$ or $x > 0$, -1, 0

(d)

x	y	y'	y''	Conclusions
-2	$\approx 1/3$	-	-	falling; concave down
-1	0	-1/3	0	falling; inflection pt.
-1/8	-1/4	0	+	"Holds water"; min.
0	0	+	0	rising; inflection pt.
1	2	+	-	rising; concave down

(e)

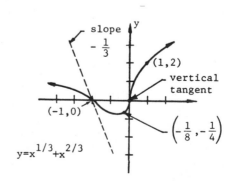

3-4   MAXIMA AND MINIMA: THEORY.

OBJECTIVE A :   Define the terms relative maximum, relative minimum, absolute maximum, and absolute minimum.

21.   A function f is said to have a relative maximum at x = a if _____ for all positive and negative values of h near _____ .

22.   A function f is said to have a _____ maximum over its domain at x = a if f(a) ≥ f(x) for all x belonging to the _____ of f .

23.   A function f is said to have a _____ minimum over its domain at x = a if f(a) ≤ f(x) for all x close to a .

24.   If f(a) ≤ f(x) for all x in the domain of f , then f is said to have a _____ _____ at x = a .

25.   Can a relative maximum also be an absolute maximum for a function f? Can a relative minimum also be an absolute minimum?

OBJECTIVE B :   Interpret correctly the theorem in this article of the text relating relative extrema at an interior point x = c of the domain a ≤ x ≤ b of a function f and the derivative $f'(c)$.

Answer questions 26-29 true or false.

26.   If $f'(c)$ = 0, then f has either a relative maximum or a relative minimum at the interior point x = c. (True or False)

27.   If f has a relative minimum at the interior point x = c, then $f'(c)$ = 0. (True or False)

28.   If f has a relative maximum or relative minimum at an end-point of the interval of definition of the function, then the lefthand (or righthand) tangent must have slope zero there. (True or False)

29.   If f has an absolute maximum at an interior point x = c and $f'(c)$ exists as a finite number, then $f'(c)$ is necessarily zero. (True or False)

30.   A point on the curve y = f(x) at which $f'(x)$ = 0 is called a _____ point of the curve. Values of x that satisfy the equation $f'(x)$ = 0 are called _____ values of the function f .

---

21. f(a) ≥ f(a + h), zero   22. absolute, domain   23. relative   24. absolute minimum

25. yes to both questions   26. False, it could have a point of inflection

27. False, the derivative may   28. False   29. True   30. stationary,
     fail to exist   critical

31.   If f is continuous over the closed interval $a \leq x \leq b$, then every point where f has a (relative or absolute) maximum or minimum must be an _____ of the interval, a point where f' _____, or an _____ point where f' equals _____.

32.   If $\frac{dy}{dx} = 0$ and $\frac{d^2y}{dx^2} > 0$, then y is a _____.

33.   If $\frac{dy}{dx} = 0$ and $\frac{d^2y}{dx^2} < 0$, then y is a _____.

34.   If $\frac{dy}{dx} = 0$ and $\frac{d^2y}{dx^2} = 0$, the second derivative test _____.

35.   Another test for a relative maximum at $x = c$ is $f'(c) = 0$ and $f'(x)$ positive for _____ and negative for _____ ; for a relative minimum at $x = c$ the conditions are $f'(c) = 0$, $f'(x)$ _____ for $x < c$ and _____ for $x > c$. If dy/dx does not change sign as x advances through c, neither a maximum nor a minimum need occur.

OBJECTIVE C :   Given a function $y = f(x)$ defined over a closed interval $a \leq x \leq b$, find the critical points of f and for each critical point, determine whether the function has a local maximum or local minimum there, or neither. If possible, find the absolute maximum and minimum values of the function on the indicated domain.

36.   Consider $y = x^{3/2}(x - 8)^{-1/2}$ over $10 \leq x \leq 16$. Now,

$y' = $ _____ $= \frac{1}{2} x^{1/2}(x - 8)^{-3/2}$ _____. Thus,

$x = $ _____ is the only critical point in the interval $10 \leq x \leq 16$. Now when $x = 12$, $y \approx 20.78$. Since $y(11) \approx 21.06$ and $y(13) \approx 20.96$, we conclude that the function has a local _____ when $x = 12$. Checking the endpoints of the interval [10,16] we determine that $y(10) \approx 22.36$ and $y(16) \approx 22.63$. Thus the absolute maximum of y occurs at $x = $ _____ and the absolute minimum of y occurs at $x = $ _____.

---

31. endpoint, does not exist, interior, 0   32. minimum

33. maximum   34. fails

35. $x < c$, $x > c$, negative, positive

36. $\frac{3}{2} x^{1/2}(x - 8)^{-1/2} - \frac{1}{2} x^{3/2}(x - 8)^{-3/2}$, 2x - 24, 12, minimum, 16, 12

3-5   MAXIMA AND MINIMA: PROBLEMS.

OBJECTIVE :   Solve a maxima or minima problem by
1. Drawing a figure, if possible, to illustrate the problem.
2. Writing an equation for the quantity y that is to be a maximum or minimum.
3. Finding and testing the critical points for a possible maximum or minimum.
4. Testing the endpoints of the interval over which y is defined (if any exist) for a possible maximum or minimum.

37.   At 9:00 A.M. ship B was 65 miles due east of another ship A. Ship B was then sailing due west at 10 miles per hour, and ship A was sailing due south at 15 miles per hour. If they continue to follow their respective courses, when will they be nearest one another and how near?

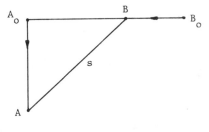

Solution. Let A_0 and B_0 denote the original positions of the ships at 9:00 A.M., and let A   and   B denote their new positions, respectively, at t hours later. This is pictured in the figure at the right. Let s denote the distance between A and B. The problem is to minimize _____ and to find the time when its minimum occurs. Since rate x time = distance, the distance covered by ship A in t hours is _____ miles, and by ship B _____ miles. The original distance between A and B is given as 65 miles, so the distance between the original position A_0 and ship B after t hours is _____ . Fill this information into the figure and then calculate the square of the distance: $s^2 =$ _____ . Differentiation of both sides of this equation with respect to t gives $2s\, ds/dt =$ _____ . Thus, $ds/dt = 0$ when $30(15t) - 20(65 - 10t)$ equals 0, or t = _____ hours. Simplifying ds/dt algebraically, we see that $ds/dt =$ _____ . Thus, ds/dt is _____ when t < 2 and _____ when t > 2.   Therefore, a relative _____ distance occurs for s at t = 2 hours. Solving for the distance s after two hours, we find

$$s^2 = (30)^2 + (\text{_____})^2 \quad \text{or} \quad s = \text{_____ miles,}$$

the distance the ships are apart at 11:00 A.M. when they are nearest each other.

---

37. s, 15t, 10t, 65 - 10t, $(15t)^2 + (65 - 10t)^2$,

$30(15t) - 20(65 - 10t)$, 2, $\dfrac{325t - 650}{s}$,

negative, positive, minimum, 45, $15\sqrt{13}$

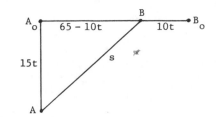

38. A company's cost function is $C(x) = 10x + 3$, and its revenue
function is $R(x) = 50x - 0.5x^2$, both in thousands of dollars
per thousand items. Find the company's maximum profit.
Solution. If $P(x)$ denotes the profit function, then
$P(x) = R(x) -$ _____ $=$ _____. The maximum profit occurs
when $P'(x) =$ _____, so $P'(x) =$ _____ $= 0$. Thus,
$x =$ _____ thousand items. Since $P''(x) =$ _____ is always
negative, this yields a maximum profit of $P(40) =$ _____
thousand dollars.

39. It is know that the population P for the fur-bearing snowshoe
hare in the Hudson Bay area will grow to $f(P) = -0.025P^2 + 4P$
in one year. If they "harvest" the amount $f(P) - P$ so the
initial population is not depleted, then the harvest is said to
be "sustained." Find the population at which the maximum
sustainable harvest occurs, and find the maximum sustainable
harvest for the snowshoe hare. Assume P is measured in
thousands.
Solution. The harvest function $H(P) = f(P) - P =$ _____.
The maximum sustainable harvest occurs when $H'(P) =$ _____, so
$H'(P) =$ _____ $= 0$, or $P =$ _____ thousand hares. This
is the population at which the maximum sustainable harvest
occurs, since $H''(P) =$ _____ is always _____. The maximum
harvest is $H(60) =$ _____ thousand animals.

40. Determine the point on the ellipse $4x^2 + 9y^2 = 36$ that is
nearest the origin.
Solution. The problem is to _____ the distance s from a
point (x,y) on the ellipse to the origin. That is, find the
minimum of $s =$ _____ subject to the auxiliary condition
that $4x^2 + 9y^2 = 36$. We can just as well minimize $s^2 = S$
since that will also minimize s. Since $S = x^2 + y^2$ is a
function of both the variables x and y, we use the equation
of the ellipse to eliminate the variable y: $y^2 = 4 - \frac{4}{9}x^2$ so
that substitution gives
$$S(x) = x^2 + y^2 = \underline{\hspace{2cm}} .$$

The minimum distance occurs when $dS/dx =$ _____, so
$(10/9)x = 0$ or $x =$ _____. Since $S''(x) > 0$ this value of
x yields a _____. When $x = 0$, $y^2 = 4$ on the ellipse
so that $(0,2)$ and $(0,-2)$ are the points on the ellipse that
are nearest to the origin.

---

38. $C(x)$, $40x - 0.5x^2 - 3$, 0, $40 - x$, 40, -1, 797

39. $-0.025P^2 + 3P$, 0, $-0.05P + 3$, 60, -0.05, negative, 90

40. minimize, $\sqrt{x^2 + y^2}$, $\frac{5}{9}x^2 + 4$, 0, 0, minimum

41.   A lighthouse is at a point A, 4 miles offshore from the
      nearest point O of a straight beach; a store is at point B,
      4 miles down the beach from O. If the lighthouse keeper can
      row 4 miles/hour and walk 5 miles/hour, find the point C
      on the beach to which the lighthouse keeper should row to get
      from the lighthouse to the store in the least possible time.

      <u>Solution</u>. The information is sketched
      in the figure at the right. From the
      diagram and the Pythagorean theorem,
      the distance from A to C is
      _____. The total time required to
      get from A to C to B is

      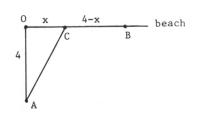

$$T = \frac{\sqrt{16 + x^2}}{4} + \left(\frac{}{5}\right),$$

      where $0 \le x \le 4$. The minimum time occurs when $dT/dx = 0$,   or

$$0 = T'(x) = \underline{}.$$

      Simplifying algebraically, $5x = $ _____ or $x^2 = $ _____   or
      $x = $ _____. However, $x = 16/3$ is outside the allowable
      range of values $0 \le x \le 4$. Therefore, the minimum must be
      taken on at one of the _____ of the interval. Checking each
      point, $T(0) = $ _____ hours and $T(4) = $ _____ hours. The
      smaller of these values occurs when $x = $ _____ so our
      conclusion is that the lighthouse keeper should row all the way
      to get to the store in the least possible time.

42.   The cost per hour of driving a ship through the water varies
      approximately as the cube of its speed in the water. Suppose a
      ship runs into a current of V miles per hour, measured
      relative to the ocean bottom. Find the total cost for the ship
      to travel M miles, and find the most economical speed of the
      ship relative to the ocean bottom.

      <u>Solution</u>. Let x denote the speed of the ship relative to the
      water. Then _____ will be its speed relative to the bottom.
      The time taken to travel M miles will be _____. The cost
      per hour in fuel will be kx^3 for some constant of proportion-
      ality k, so the total cost function is given by

$$C(x) = \underline{}.$$

      To find the most economical speed, we want to minimize the cost

$$C'(x) = \underline{}.$$

      The minimum cost occurs when $dC/dx = 0$,   or $kMx^2[3(x - V) - x]$
      $= 0$. Thus, $x = $ _____ or $x = $ _____. Since $x = 0$ is
      ruled out if the ship moves, and since $C(x) \to +\infty$ as
      $x \to V^+$, we see that $x = 1.5V$ must provide the <u>minimum</u> cost.

---

41. $\sqrt{16 + x^2}$, 4 - x, $\frac{1}{8}(16 + x^2)^{-1/2}(2x) - \frac{1}{5}$, $4\sqrt{16 + x^2}$, $\frac{256}{9}$, $\frac{16}{3}$, endpoints, 1.8, $\sqrt{2}$, 4

42. x - V, M/(x - V), $\frac{kMx^3}{x - V}$, $\frac{(x - V)3kMx^2 - kMx^3}{(x - V)^2}$, 0, 1.5V

3-6 RELATED RATES OF CHANGE.

In this article we consider problems that ask us to find the rate at which some variable quantity changes when we know the rate at which another quantity related to it changes. Examples abound for problems of this sort. For instance, the rate of production of a certain commodity may depend upon its rate of sales; the rate of increase or decrease in the water level of a dam or reservoir is essential information to a public utility serving the demands of a growing population; the rate at which oil may be spreading on the sea surface from a stricken tanker depends on the rate at which it may be leaking; and so forth.

OBJECTIVE : Solve a related rates problem by
1. Drawing a figure, if possible, to illustrate the problem. Use t for time and assume that all variables are differentiable functions of t.
2. Writing down any additional numerical information.
3. Writing down what you are asked to find.
4. Writing an equation that relates the variables. You may have to combine several equations to get a single equation that relates the variable whose rate you want to the variable whose rate you know.
5. Differentiating to find the desired rate of change.

43. A plane flying at 1 mile altitude is 2 miles distant from an observer, measured along the ground, and flying directly away from the observer at 400 mph. How fast is the angle of elevation changing?

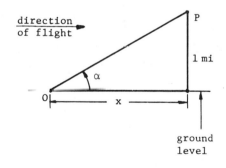

Solution. Let 0 denote the position of the observer at a distance x units (measured along the ground) from the plane P as shown in the figure at the right. Let α denote the angle of elevation. We are asked to find the rate _____. The angle α satisfies the equation tan α = _____. This equation holds for all time t. Differentiating both sides of the equation with respect to t yields,

$$\sec^2\alpha \cdot (\underline{\hspace{1cm}}) = -\frac{1}{x^2} \cdot (\underline{\hspace{1cm}}),$$

or, since $\sec^2\alpha = 1 + \tan^2\alpha = \underline{\hspace{1cm}}$, we solve to find
$$\frac{d\alpha}{dt} = (\underline{\hspace{1cm}}) \frac{dx}{dt} .$$

Thus, when x = 2 and dx/dt = 400, $\frac{d\alpha}{dt}$ = _____ radians per hour, or _____ rad/sec, or _____ deg/sec. Notice that the angle of elevation is decreasing because dα/dt is _____.

---

43. dα/dt, 1/x, dα/dt, dx/dt, $1 + \frac{1}{x^2}$, $-\frac{1}{x^2}\left(\frac{x^2}{1+x^2}\right)$, -80, $\frac{-80}{3600}$, (approx) -1.3, negative

44.   A trough 10 ft long has a cross section that is an isosceles
      triangle 3 ft deep and 8 ft across. If water flows in at
      the rate of 2 cu ft/min, how fast is the surface rising when
      the water is 2 ft deep?

      <u>Solution</u>. A cross section of the trough
      is shown in the figure at the right. In
      the figure h denotes the depth of the
      water and b its width across the trough
      at any instant t. Thus, b and h are
      both functions of _____. At any instant of time the volume of
      water in the trough is given by the formula,

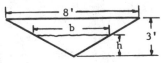

$$V = 10 \ (\underline{\hspace{2cm}}) \ .$$

      We are given the rate dV/dt = 2 and we are asked to find the
      rate _____ when h = 2. Since the formula for V involves
      both the variables b and h, we need to write down a formula
      relating these variables. From the geometry of similar
      triangles in the figure we have

$$\frac{b}{\underline{\hspace{1cm}}} = \frac{8}{3} \quad \text{or,} \quad b = \underline{\hspace{2cm}} \ .$$

      Substitution into the formula for V gives V = _____.
      Differentiation of both sides of this last equation with
      respect to t yields,

$$\frac{dV}{dt} = \underline{\hspace{3cm}} \ .$$

      Solving for dh/dt when dV/dt = 2 and h = 2 gives

$$dh/dt = \underline{\hspace{2.5cm}} \ \text{ft/min.}$$

45.   A walk is perpendicular to a long wall, and a man strolls along
      it away from the wall at the rate of 3 ft/sec. There is a
      light 8 ft from the walk and 24 ft from the wall. How fast
      is his shadow moving along the wall when he is 20 ft from the
      wall?

      <u>Solution</u>. The situation is pictured
      in the figure at the right. Here M
      denotes the position of the man at a
      distance x units from the wall, S
      denotes the position of his shadow on
      the wall at a distance y units from
      where the wall and the walk intersect,
      and L denotes the position of the
      light. We are given that _____ = 3
      and are asked to find _____ when
      x = 20. From similar triangles we can

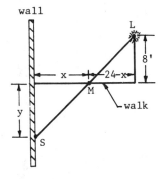

      establish a relationship between the variables x and y,

$$\frac{x}{y} = \frac{\underline{\hspace{1cm}}}{8} \ , \quad \text{or} \quad 8x = \underline{\hspace{2cm}} \ .$$

      This equation is valid at any instant of time t. Differen-
      tiating both sides with respect to t yields

---

44. t, $\frac{1}{2}$ bh, dh/dt, h, $\frac{8}{3}$ h, $\frac{40}{3}$ h², $\frac{80}{3}$ h $\frac{dh}{dt}$, $\frac{3}{80}$

$$8 \frac{dx}{dt} = (\underline{\hspace{1cm}}) \frac{dy}{dt} - \underline{\hspace{1cm}}.$$

When $x = 20$, $y = \frac{8(20)}{\underline{\hspace{0.5cm}}} = \underline{\hspace{2cm}}$, so substitution into the previous derivative equation yields

$$8 \cdot 3 = \underline{\hspace{1.5cm}} - 120, \quad \text{or} \quad \frac{dy}{dt} = \underline{\hspace{1cm}} \text{ ft/sec}.$$

3-7 THE MEAN VALUE THEOREM.

OBJECTIVE A : Apply Rolle's Theorem to show that a given equation $f(x) = 0$ has exactly one solution in the specified interval $a \leq x \leq b$.

46. Suppose $y = f(x)$ and its first derivative $f'(x)$ are continuous over $a \leq x \leq b$. If $f(a)$ and $f(b)$ have opposite sign, then according to the Intermediate Value Theorem there is at least one point c satisfying $a < c < b$ and $f(c) = \underline{\hspace{1.5cm}}$.

47. Suppose there is another point d satisfying $a < d < b$ and $f(d) = 0$. Then, according to Rolle's Theorem, there is a point between c and d for which $\underline{\hspace{1.5cm}}$ is zero. Thus, if $f'(x)$ is different from zero for all values of x between a and b, there is exactly $\underline{\hspace{1.5cm}}$ solution to the equation $f(x) = 0$ in the interval $\underline{\hspace{1.5cm}}$.

48. Consider the equation $x^3 + 2x^2 + 5x - 6 = 0$ for $0 \leq x \leq 5$. When $x = 0$ the value of the left side is $\underline{\hspace{1.5cm}}$; and when $x = 5$, the value is $\underline{\hspace{1.5cm}}$. These values differ in sign. Calculating the derivative, we have

$$\frac{d}{dx}\left(x^3 + 2x^2 + 5x - 6\right) = \underline{\hspace{3cm}},$$

and this is always $\underline{\hspace{1.5cm}}$ for $0 < x < 5$. Therefore, we conclude from Problems 46 and 47 that there is exactly one solution to the equation somewhere between $x = \underline{\hspace{1.5cm}}$ and $x = \underline{\hspace{1.5cm}}$. We could in fact use Newton's method of Article 2-9 to locate this solution.

OBJECTIVE B : Given a function $y = f(x)$ satisfying the hypotheses of the Mean Value Theorem for $a \leq x \leq b$, use the theorem to find a number c satisfying the conclusion of the theorem.

49. The hypotheses of the Mean Value Theorem are that f is $\underline{\hspace{3cm}}$ over the closed interval $a \leq x \leq b$ and $\underline{\hspace{3.5cm}}$ over the open interval $\underline{\hspace{2cm}}$.

---

45. dx/dt, dy/dt, $24 - x$, $(24 - x)y$, $24 - x$, $y\frac{dx}{dt}$, $24 - 20$, 40, $4\frac{dy}{dt}$, 36 46. 0

47. $f'(x)$, one, $a \leq x \leq b$ 48. -6, 194, $3x^2 + 4x + 5$, positive, 0, 5

49. continuous, differentiable, $a < x < b$

50.   The conclusion of the Mean Value Theorem is that there is at
least one point   c   in the open interval _____ satisfying
_____.   A geometric interpretation of the conclusion is
that the slope of the curve $y = f(x)$ when $x = c$ is the same
as the slope of the _____ joining the endpoints $(a, f(a))$ and
_____ of the curve.

51.   Let $f(x) = 3x^2 + 4x - 3$ over $1 \leq x \leq 3$.   Then $f'(x) =$ _____,
so that   f   and f' satisfy the hypotheses of the Mean Value
Theorem.   To find a value for   c,   the equation $f(b) - f(a) =$
$f'(c)(b - a)$ becomes $f(3) - f(1) = f'(c)($_____$)$,   or
$36 - 4 = 2($_____$)$.   Solving for   c   gives $c =$ _____.

52.   Does the Mean Value Theorem apply to the function $f(x) = |x|$
in the interval $[-2, 1]$?
No, because the derivative $f'(x)$ is not defined for
$x =$ _____ so the function   f   is not _____ over the
open interval _____ as required by the hypotheses.

$\boxed{\text{OBJECTIVE C}}$:   Use the approximation $f(b) \approx f(a) + (b - a)f'(a)$ to
make reasonable estimates for the value of $y = f(x)$
when $x = b$.

53.   Estimate $\sin 29°$.
Here $f(x) = \sin x$,   but we must use radian measure.   Thus,
$b = 29° =$ _____ rad   and $a =$ _____ rad.   The approximation
then becomes,

$$\sin 29° \approx \sin \frac{\pi}{6} + (\underline{\hspace{1cm}}) = \frac{1}{2} - (\underline{\hspace{1cm}}) = .48489.$$
(The error is less than 10^{-4} from the true value.)

3-8   INDETERMINATE FORMS AND L'HÔPITAL'S RULE.

$\boxed{\text{OBJECTIVE A}}$:   Evaluate the limits of indeterminate forms using
l'Hôpital's rule, whenever applicable.

54.   The three hypotheses necessary to apply l'Hôpital's rule

$\lim_{x \to a} \dfrac{f(x)}{g(x)} = \lim_{x \to a} \dfrac{f'(x)}{g'(x)}$ in the indeterminate case $0/0$ are:

(a)   f   and   g   are both _____ over an open interval   I
containing the point $x = a$,   except possibly at _____;

(b) $f(a) = g(a) =$ _____;   and

(c) $g'(x)$ _____ over   I,   except possibly when _____.

---

50. (a,b), $f(b) - f(a) = f'(c)(b - a)$, chord, $(b, f(b))$ 51. $6x + 4$, 2, $6c + 4$, 2

52. 0, differentiable, $(-2, 1)$ 53. $\dfrac{29\pi}{180}$, $\dfrac{30\pi}{180}$, $-\dfrac{\pi}{180} \cos \dfrac{\pi}{6}$, $\dfrac{\pi}{180}\left(\dfrac{\sqrt{3}}{2}\right)$

54. differentiable, a, 0, $\neq 0$, $x = a$

55. $\lim\limits_{u \to 1} \dfrac{u^4 - 1}{\sqrt{u} - 1}$ is of the form _____. Applying l'Hôpital's rule,

$$\lim\limits_{u \to 1} \dfrac{u^4 - 1}{\sqrt{u} - 1} = \lim\limits_{u \to 1} \underline{\hspace{3cm}} = \dfrac{4}{\underline{\hspace{1cm}}} = \underline{\hspace{2cm}} \; .$$

56. $\lim\limits_{x \to \infty} \dfrac{\sin \frac{7}{x^2}}{\frac{3}{x^2}}$ is of the form _____ .

Since $d/dx \, (3/x^2) = -6/x^3$ is never _____, l'Hôpital's rule applies. Thus,

$$\lim\limits_{x \to \infty} \dfrac{\sin \frac{7}{x^2}}{\frac{3}{x^2}} = \lim\limits_{x \to \infty} \underline{\hspace{3cm}} = \lim\limits_{x \to \infty} \left(\underline{\hspace{1cm}} \right) \cos \dfrac{7}{x^2} = \underline{\hspace{2cm}} \; .$$

57. $\lim\limits_{x \to \infty} \dfrac{x^2 - 4x + 200}{5x^4 - 7x^2 + 21}$ is of the form _____. Applying l'Hôpital's rule,

$$\lim\limits_{x \to \infty} \dfrac{x^2 - 4x + 200}{5x^4 - 7x^2 + 21} = \lim\limits_{x \to \infty} \underline{\hspace{2cm}} \; [\text{still} \underline{\hspace{2cm}}]$$

$$= \lim\limits_{x \to \infty} \dfrac{2}{60x} = \underline{\hspace{2cm}} \; .$$

58. $\lim\limits_{x \to \frac{\pi}{2}^-} (\sec x - \tan x)$ is of the form ___. We employ trigonometric identities to convert the form to another indeterminate 0/0 form, and apply l'Hôpital's rule:

$$\lim\limits_{x \to \frac{\pi}{2}^-} (\sec x - \tan x) = \lim\limits_{x \to \frac{\pi}{2}^-} \left(\dfrac{\underline{\hspace{1.5cm}}}{\cos x} \right) \qquad \left[\dfrac{0}{0}\right]$$

$$= \lim\limits_{x \to \frac{\pi}{2}^-} \underline{\hspace{2cm}} = \underline{\hspace{1.5cm}} \; .$$

59. $\lim\limits_{x \to \frac{\pi}{2}^+} (\sec x - \tan x)$ is of the form _____. As in Problem 58,

$$\lim\limits_{x \to \frac{\pi}{2}^+} (\sec x - \tan x) = \lim\limits_{x \to \frac{\pi}{2}^+} \left(\underline{\hspace{2cm}} \right) \qquad \left[\dfrac{0}{0}\right]$$

$$= \lim\limits_{x \to \frac{\pi}{2}^+} \underline{\hspace{2cm}} = \underline{\hspace{1.5cm}} \; .$$

We conclude that $\lim\limits_{x \to \frac{\pi}{2}} (\sec x - \tan x) = \underline{\hspace{2cm}}$.

---

55. $0/0$, $\dfrac{4u^3}{1/2\sqrt{u}}$, $\dfrac{1}{2}$, 8

56. $0/0$, zero, $\dfrac{\frac{-14}{x^3} \cos \frac{7}{x^2}}{-\frac{6}{x^3}}$, $\dfrac{14}{6}$, $\dfrac{7}{3}$

57. ∞/∞, $\dfrac{2x - 4}{20x^3 - 14x}$, $\dfrac{\infty}{\infty}$, 0

58. $\infty - \infty$, $1 - \sin x$, $\dfrac{-\cos x}{-\sin x}$, 0

59. $-\infty - (-\infty)$ or $\infty - \infty$ again, $\dfrac{1 - \sin x}{\cos x}$, $\dfrac{-\cos x}{-\sin x}$, 0, 0

60. $\lim\limits_{x \to +\infty} x^2 \sin \frac{1}{x}$ is of the form _____. We can reduce this to a

form   0/0   by algebraic manipulation and then apply l'Hôpital's
rule:

$$\lim_{x \to +\infty} x^2 \sin \frac{1}{x} = \lim_{x \to +\infty} \frac{\sin \frac{1}{x}}{\underline{\qquad}} \quad \left[\frac{0}{0}\right] = \lim_{x \to +\infty} \underline{\qquad}$$

$$= \lim_{x \to +\infty} \left(\underline{\qquad}\right) \cos \frac{1}{x} = \underline{\qquad} \; .$$

There is no finite limit.

61.   Which argument is correct, (a) or (b)? Explain.

(a) $\lim\limits_{x \to 2} \dfrac{x^3 - x^2 - x - 2}{x^3 - 3x^2 + 3x - 2} = \lim\limits_{x \to 2} \dfrac{3x^2 - 2x - 1}{3x^2 - 6x + 3} = \lim\limits_{x \to 2} \dfrac{6x - 2}{6x - 6}$

$$= \lim_{x \to 2} \frac{6}{6} = 1$$

(b) $\lim\limits_{x \to 2} \dfrac{x^3 - x^2 - x - 2}{x^3 - 3x^2 + 3x - 2} = \lim\limits_{x \to 2} \dfrac{3x^2 - 2x - 1}{3x^2 - 6x + 3} = \dfrac{12 - 4 - 1}{12 - 12 + 3} = \dfrac{7}{3}.$

3-9   QUADRATIC APPROXIMATIONS AND APPROXIMATION ERRORS: EXTENDING THE MEAN VALUE THEOREM.

OBJECTIVE :   For a given function $y = f(x)$ satisfying the hypotheses
of the Extended Mean Value Theorem, find a quadratic
approximation for f near a specified value $x = a$.

62.   Let $f(x) = \tan x$.   Then $f'(x) =$ _____ and $f''(x) =$ _____.
Hence, $f(\pi/4) =$ _____, $f'(\pi/4) = 2$,   and $f''(\pi/4) =$ _____.
Therefore, a quadratic approximation to f near $x = \pi/4$ is
given by

$$f(x) = 1 + \underline{\qquad} \left(x - \frac{\pi}{4}\right) + \underline{\qquad\qquad} \; .$$

63.   Consider the polynomial $f(x) = 5x^4 - 3x^3 + x^2 - 7$.   Take $a = 2$
so we will express $f(x)$ in powers of $x - 2$ rather than
powers of x. We need to calculate the first four derivatives
of f and evaluate them when $x =$ _____.   Thus,
$f'(x) =$ _____   so $f'(2) =$ _____ ,
$f''(x) =$ _____   so $f''(2) =$ _____ ,
$f^{(3)}(x) =$ _____   so $f^{(3)}(2) =$ _____ ,
$f^{(4)}(x) =$ _____   so $f^{(4)}(2) =$ _____ .
Observe that all higher derivatives $f^{(n)}(x)$, $n \geq 5$,   are
identically _____.   Therefore, by the Extended Mean Value
Theorem, $f(x) =$ _____ + _____$(x - 2)$ + ____$(x - 2)^2$ + ____$(x - 2)^3$
+ ____$(x - 2)^4 =$ _____ .

---

60. $\infty \cdot 0$, $\frac{1}{x^2}$, $-\frac{1}{x^2} \cos \frac{1}{x} \Big/ -\frac{2}{x^3}$, $\frac{x}{2}$, $+\infty$ 61.   (b) is correct: l'Hôpital's rule does not apply to the

computation of either $\lim\limits_{x \to 2} \dfrac{3x^2 - 2x - 1}{3x^2 - 6x + 3}$ or $\lim\limits_{x \to 2} \dfrac{6x - 2}{6x - 6}$ because the numerators and denominators

have finite, nonzero limits   62. $\sec^2 x$, $2 \sec x \tan x$, 1, $2\sqrt{2}$, 2, $\sqrt{2}\left(x - \frac{\pi}{4}\right)^2$

63. $(x - 2)$, 2, $20x^3 - 9x^2 + 2x$, 128, $60x^2 - 18x + 2$, 206, $120x - 18$, 222, 120, 120, zero, 53,

128, $\dfrac{206}{2!}$, $\dfrac{222}{3!}$, $\dfrac{120}{4!}$, $f(x) = 53 + 128(x - 2) + 103(x - 2)^2 + 37(x - 2)^3 + 5(x - 2)^4$

CHAPTER 3   SELF-TEST

In Problems 1-3, sketch the curves. Find the intervals of values of x for which the curve is increasing, decreasing, concave upward, and concave downward. Locate all asymptotes.

1. $y = \dfrac{x}{\sqrt{1 + x^2}}$
2. $y = \dfrac{4x}{x^2 + 1}$
3. $y = 1 - (x + 1)^{1/3}$

4. A photographer is televising a 100-yard dash from a position 10 yards from the track in line with the finish line. When the runners are 10 yards from the finish line, the camera is turning at the rate 3/5 rad/sec. How fast are the runners moving then?

5. A swimming pool is 40 ft long, 20 ft wide, 8 ft deep at the deep end, and 3 ft deep at the shallow end, the bottom being rectangular. If the pool is filled by pumping water into it at the rate of 40 cu. ft/min, how fast is the water level rising when it is

   (a) 3 ft deep at the deep end?
   (b) 6 ft deep at the deep end?

6. A guy wire is to pass from the top of a pole 36 ft high to an anchorage on the ground 27 ft from the base of the pole. One end of the wire is made fast to the anchorage, and a man climbs the pole with the wire, keeping it taut. If he climbs 2 ft/sec, how fast is he playing out the wire when he reaches the top of the pole?

7. Find the absolute maximum and minimum values (if they exist) of $f(x) = x^3 - x^2 - x + 2$ over the interval $0 \le x < 2$.

8. Suppose a company can sell x items per week at a price P = 200 - 0.01x cents, and that it costs C = 50x + 20,000 cents to produce the x items. How much should the company charge per item in order to maximize its profits?

9. The weight W (lbs/sec) of flue gas passing up a chimney at different temperatures T is represented by

   $$W = A(T - T_0)(1 + \alpha T)^{-2} ,$$

   where A is a positive constant, T the absolute temperature of the hot gases passing up the chimney, T_0 the temperature of the outside air (all in °C), and $\alpha = 1/273$ is the coefficient of expansion of the gas. For a given $T_0 = 15$°C, find the temperature T at which the greatest amount of gas will pass up the chimney.

10. Apply Rolle's Theorem to show that the equation $\cos x = \sqrt{x}$, $x \ge 0$, has exactly one real solution.

11. Find all numbers c which satisfy the conclusion of the Mean Value Theorem for $f(x) = 1 + 2x^2$ over $-1 \le x \le 1$.

12. Let $f(x) = \frac{1}{x}$. Show that there is no c in the interval $-1 < x < 2$ such that $f'(c) = \frac{f(2) - f(-1)}{2 - (-1)}$. Explain why this does <u>not</u> contradict the Mean Value Theorem.

13. Find a reasonable estimate to $\sqrt[3]{25}$.

Evaluate the limits in Problems 14-17.

14. $\displaystyle\lim_{x \to 0} \frac{\sin x - x \cos x}{x^3}$

15. $\displaystyle\lim_{x \to \infty} \frac{x^2 - 5}{2x^2 + 3x}$

16. $\displaystyle\lim_{x \to \infty} \frac{\sin(3/x)}{2/x}$

17. $\displaystyle\lim_{x \to \frac{\pi^-}{2}} \left(x \tan x - \frac{\pi}{2} \sec x \right)$

18. Find a quadratic approximation for the function $f(x) = \sec x$ near $x = 0$.

SOLUTIONS TO CHAPTER 3 SELF-TEST

1. $y = \dfrac{x}{(1 + x^2)^{1/2}}$

   $y' = \dfrac{(1 + x^2)^{1/2} - x \cdot \frac{1}{2}(1 + x^2)^{-1/2} 2x}{1 + x^2} = \dfrac{1}{(1 + x^2)^{3/2}}$, and $y'' = \dfrac{-3x}{(1 + x^2)^{5/2}}$.

   Note that $y = \dfrac{1}{(1/x^2 + 1)^{1/2}}$ for $x \geq 0$ $\left(\text{since } \sqrt{x^2} = |x| \right)$

   and that $y = \dfrac{-1}{(1/x^2 + 1)^{1/2}}$ for $x < 0$. Thus $\displaystyle\lim_{x \to \infty} y = 1$ and

   $\displaystyle\lim_{x \to -\infty} y = -1$. Hence the lines $y = 1$ and $y = -1$ are <u>horizontal</u> <u>asymptotes</u>. Since y' exists for all x and is never zero, there are <u>no</u> <u>critical</u> <u>points</u>. At $x = 0$, $y'' = 0$ so that $x = 0$ is a <u>point</u> <u>of inflection</u> where the graph has slope 1. On $(-\infty, 0)$ $y'' > 0$ and the function is concave upward; on $(0, \infty)$ it is concave downward. Since $y' > 0$ for all x, the graph is everywhere an increasing function of x. This information yields the graph sketched at the right.

   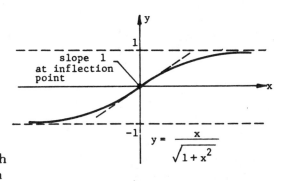

2. Since $\displaystyle\lim_{x \to \pm\infty} y = \lim_{x \to \pm\infty} \frac{4/x}{1 + 1/x^2} = 0$, the x-axis is a <u>horizontal</u> <u>asymptote</u>. Next, $y' = \dfrac{4(1 - x^2)}{(x^2 + 1)^2}$ and $y'' = \dfrac{8x(x^2 - 3)}{(x^2 + 1)^3}$.

   Hence, $y' = 0$ implies $x = \pm 1$. Since $y'' > 0$ at $x = -1$ and $y'' < 0$ at $x = 1$, it follows from the second derivative test that $y(-1) = -2$ is a <u>relative</u> <u>minimum</u> and $y(1) = 2$ is a <u>relative</u> <u>maximum</u>.

Next, $y'' = 0$ when $x = 0$, $-\sqrt{3}$, and $\sqrt{3}$ so that these values for x are <u>points of inflection</u>. Moreover, for

$x < -\sqrt{3}$, $y'' < 0$ and the graph of y is concave downward;

$-\sqrt{3} < x < 0$, $y'' > 0$ and the graph of y is concave upward;

$0 < x < \sqrt{3}$, $y'' < 0$ and the graph of y is concave downward;

$x > \sqrt{3}$, $y'' > 0$ and the graph of y is concave upward.

Note that at $x = 0$, $y' = 4$. The graph of y is sketched at the right. Note the symmetry about the origin.

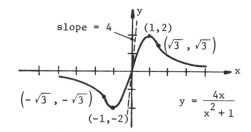

3. $y = 1 - (x + 1)^{1/3}$, $y' = -\frac{1}{3}(x + 1)^{-2/3}$, and $y'' = \frac{2}{9}(x + 1)^{-5/3}$.

The derivative y' does not exist when $x = -1$, although the curve y is continuous at $x = -1$. Since

$$\lim_{x \to -1} \frac{dx}{dy} = \lim_{x \to -1} -3(x + 1)^{2/3} = 0,$$

the graph has a vertical tangent at $x = -1$. Since $y' < 0$ for all $x \neq -1$, the curve is everywhere decreasing.

We note that y'' is never zero. However, y'' fails to exist at $x = -1$. When $x < -1$, $y'' < 0$ and the curve is concave downward; when $x > -1$, $y'' > 0$ and the curve is concave upward. Therefore $x = -1$ is a point of inflection. The graph is sketched in the figure at the right.

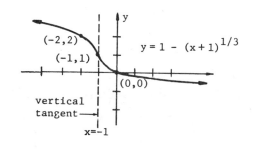

4. The situation is pictured in the figure at the right. Thus,

$\tan \theta = \frac{x}{10}$, or $x = \tan \theta$.

$\frac{dx}{dt} = 10 \sec^2 \theta \frac{d\theta}{dt}$. Now, when $x = 10$ yds, $\theta = \frac{\pi}{4}$, and $\frac{d\theta}{dt} = \frac{3}{5}$ rad/sec. Hence,

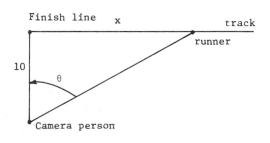

$$\frac{dx}{dt}\bigg|_{x=10} = 10\left(\sec^2 \frac{\pi}{4}\right)\left(\frac{3}{5}\right)$$

$$= 10\left(\sqrt{2}\right)^2 \left(\frac{3}{5}\right) = 12 \quad \text{yd/sec.}$$

5. A vertical cross-section of the
    pool is pictured in the figure
    at the right: y denotes the
    depth of the water at any time
    t, and x denotes the horizontal
    length of the water in the bottom
    of the pool.

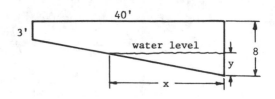

(a) When $y < 5'$, we have from the geometry of similar

triangles in the figure that, $\frac{x}{40} = \frac{y}{5}$ or $x = 8y$. The

volume of water in the pool is given by $V = \frac{1}{2}x \cdot y \cdot 20$

$= 80y^2$. Hence $\frac{dV}{dt} = 160y \frac{dy}{dt}$, and since $\frac{dV}{dt} = 40$ is given,

solving for dy/dt yields $\frac{dy}{dt}\Big|_{y=3} = \frac{40}{160(3)} = \frac{1}{12}$ ft/min.

(b) When $y > 5'$, the total volume of water is given by
    $V = \frac{1}{2}(40)(5)(20) + (40)(20)(y - 5) = 800y - 2000$. Hence,

$\frac{dV}{dt} = 800 \frac{dy}{dt}$, and since $\frac{dV}{dt} = 40$ is given, solving for

dy/dt yields $\frac{dy}{dt}\Big|_{y=6} = \frac{40}{800} = \frac{1}{20}$ ft/min.

6. The situation is pictured at the
    right, where h is the height
    of the man above the ground and
    ℓ is the length of guy wire
    played out at any instant of time
    t. We want to find $\frac{d\ell}{dt}$ when

    h = 36. Now, $\ell^2 = h^2 + (27)^2$,
    and differentiation with respect
    to t gives $2\ell \frac{d\ell}{dt} = 2h \frac{dh}{dt}$.
    When h = 36, we have
    $\ell = \sqrt{(36)^2 + (27)^2} = 9\sqrt{4^2 + 3^2} = 45$.
    Thus, for dh/dt = 2,

$$\frac{d\ell}{dt}\Big|_{h=36} = \frac{h}{\ell} \frac{dh}{dt}\Big|_{h=36} = \frac{36}{45} \cdot 2 = \frac{72}{45} = \frac{8}{5} \text{ft/sec.}$$

7. $f(x) = x^3 - x^2 - x + 2$ for $0 \leq x < 2$, and
    $f'(x) = 3x^2 - 2x - 1 = (3x + 1)(x - 1)$. Thus, $f'(x) = 0$ implies
    $x = -\frac{1}{3}$ or $x = 1$. Then x = 1 is the only critical point in
    the interval [0,2). Next note that $f'(x) < 0$ in [0,1) so f
    is decreasing to the left of x = 1, and $f'(x) > 0$ in (1,2)
    so f is increasing to the right of x = 1. Also, f(0) = 2,
    f(1) = 1 and f(2) = 4. Since x = 2 is not in the interval,
    there is no absolute maximum. The absolute minimum value is
    f(1) = 1 (which is also a relative minimum).

8. Let Q denote the profit function. Then, $Q(x) = xP - C$
    $= 150x - 0.01x^2 - 20,000$. The maximum occurs when dQ/dx = 0, or
    $150 - .02x = 0$; thus x = 7500 items. Since $d^2Q/dx^2 = -.02 < 0$,

this provides a <u>maximum</u> profit. The price per item is then given by $P(7500) = 200 - (.01)(7500) = 125$ cents, the price required to obtain the maximum profit $Q(7500) = \$5,425.00$.

9. We want to maximize the weight function W. Now,

$$W = A(T - T_o)(1 + \alpha T)^{-2}$$

$$\frac{dW}{dT} = A(1 + \alpha T)^{-2} - 2A\alpha(T - T_o)(1 + \alpha T)^{-3}.$$

Setting $dW/dt = 0$, and simplifying algebraically, gives $(1 + \alpha T) - 2\alpha(T - T_o) = 0$, or $T = (1 + 2\alpha T_o)/\alpha$. Thus, for $T_o = 15°C$ and $\alpha = 1/273$ as given,

$$T = \tfrac{1}{\alpha} + 2T_o = 273 + 30 = 303°C .$$

Since $\frac{dW}{dT} > 0$ if $T < 303$, and $\frac{dW}{dT} < 0$ if $T > 303$, it is clear that $T = 303°$ provides an absolute maximum for W.

10. Let $f(x) = \cos x - \sqrt{x}$. Since $|\cos x| \leq 1$, we see that $f(x) < 0$ if $x > 1$. Thus, the only possible root must lie within the interval $[0,1]$. Now, $f(0) = 1$ and $f(\tfrac{\pi}{2}) = -\sqrt{\tfrac{\pi}{2}}$ so the Intermediate Value Theorem guarantees a root in the interval $\left[0,\tfrac{\pi}{2}\right]$: we know in fact that the root must lie in $[0,1]$. Calculating the derivative, $f'(x) = -\sin x - \frac{1}{2\sqrt{x}}$, we see that f' is negative in the interval $(0,1)$. Since f' is different from zero for all values of x between 0 and 1, we conclude there is exactly one real root to the equation $f(x) = 0$ for $x \geq 0$.

11. $f(-1) = 3$ and $f(1) = 3$; $f'(x) = 4x$. Hence $\frac{f(1) - f(-1)}{1 - (-1)} = f'(c)$ translates into $0 = 4c$, or $c = 0$.

12. $\frac{f(2) - f(-1)}{2 - (-1)} = \frac{\tfrac{1}{2} - (-1)}{3} = \tfrac{1}{2}$ and $f'(c) = -\tfrac{1}{c^2}$. Since $-\tfrac{1}{c^2} = \tfrac{1}{2}$ is impossible to solve for real values of c, there is no number c in the interval $(-1,2)$ satisfying the conclusion of the Mean Value Theorem. However, this does not contradict the Theorem because $f(x) = 1/x$ is not continuous over the closed interval $[-1,2]$: it fails to be continuous at $x = 0$. Thus the hypotheses of the theorem are not satisfied.

13. Let $f(x) = x^{1/3}$. Then by the Mean Value Theorem, $f(25) \approx f(27) + (25 - 27)f'(27)$ or,

$$\sqrt[3]{25} \approx \sqrt[3]{27} + (-2)\frac{1}{3(27)^{2/3}} = 3 - \frac{2}{3 \cdot 9} = \frac{81 - 2}{27} = \frac{79}{27} \approx 2.926.$$

(A calculator gives $\sqrt[3]{25} \approx 2.924017738$.)

rule,

$$\lim_{x \to 0} \frac{\sin x - x \cos x}{x^3} = \lim_{x \to 0} \frac{\cos x - \cos x + x \sin x}{3x^2}$$

$$= \lim_{x \to 0} \frac{\sin x}{3x} = \frac{1}{3} \lim_{x \to 0} \frac{\sin x}{x} = \frac{1}{3}.$$

15. $\lim\limits_{x \to \infty} \dfrac{x^2 - 5}{2x^2 + 3x}$ is of the form ∞/∞. Applying l'Hôpital's rule,

$$\lim_{x \to \infty} \frac{x^2 - 5}{2x^2 + 3x} = \lim_{x \to \infty} \frac{2x}{4x + 3} \quad [\text{still} \quad \infty/\infty] = \lim_{x \to \infty} \frac{2}{4} = \frac{1}{2}.$$

16. $\lim\limits_{x \to \infty} \dfrac{\sin(3/x)}{2/x}$ is of the form $0/0$. Applying l'Hôpital's rule,

$$\lim_{x \to \infty} \frac{\sin(3/x)}{2/x} = \lim_{x \to \infty} \frac{(-3/x^2) \cos(3/x)}{-2/x^2} = \lim_{x \to \infty} \frac{3}{2} \cos \frac{3}{x} = \frac{3}{2} \cos 0$$

$$= \frac{3}{2}.$$

17. $\lim\limits_{x \to \frac{\pi}{2}^-} (x \tan x - \frac{\pi}{2} \sec x)$ is of the form $\infty - \infty$. However,

$$\lim_{x \to \frac{\pi}{2}^-} (x \tan x - \frac{\pi}{2} \sec x) = \lim_{x \to \frac{\pi}{2}^-} \frac{x \sin x - \frac{\pi}{2}}{\cos x} \quad \text{is of the form} \quad 0/0.$$

Applying l'Hôpital's rule,

$$\lim_{x \to \frac{\pi}{2}^-} \frac{x \sin x - \frac{\pi}{2}}{\cos x} = \lim_{x \to \frac{\pi}{2}^-} \frac{\sin x + x \cos x}{- \sin x} = \frac{1 + \left(\frac{\pi}{2}\right)(0)}{(-1)} = -1.$$

18. With $f(x) = \sec x$ and $a = 0$, $f(0) = 1$. Also,

$f'(x) = \sec x \tan x$ $\qquad\qquad$ $f'(0) = 0$

$f''(x) = \sec x \tan^2 x + \sec^3 x$ $\qquad$ $f''(0) = 1$

Thus,

$$Q(x) = f(a) + f'(a)(x - a) + \frac{1}{2} f''(a)(x - a)^2$$

$$= 1 + 0(x - 0) + \frac{1}{2}(x - 0)^2$$

$$= 1 + \frac{1}{2}x^2$$

The quadratic approximation of $\sec x$ near $x = 0$ is
$\sec x \approx 1 + \frac{1}{2}x^2$.

CHAPTER 4 INTEGRATION

INTRODUCTION.

The notion of the inverse of an operation implies an "undoing" or reversal of the operation. That is, if we first perform an operation and then perform its inverse, we return to the original state. In arithmetic, for example, subtraction is the inverse operation of addition: if we begin with the number x and add 3 we obtain $x + 3$; subtracting 3 from $x + 3$ brings us back to the original number x. In the next several sections the idea of "undoing" the differentiation process will be studied. This undoing process is commonly termed <u>indefinite</u> integration or <u>antidifferentiation</u>. The significance of the indefinite integration process will be revealed by an important theorem and by the development further on in this chapter.

4-1 INDEFINITE INTEGRALS.

Let u and v denote differentiable functions of some independent variable (say of x), and suppose a, n, and C are constants. Then the four basic integration formulas of this section are as follows:

1. $\int \frac{du}{dx} \, dx =$ _____

2. $\int au(x) \, dx =$ _____

3. $\int [u(x) + v(x)] \, dx =$ _____

4. $\int u^n \frac{du}{dx} =$ _____ , $n \neq$ _____

$\boxed{\text{OBJECTIVE A}}$:   Find indefinite integrals of elementary functions using the formulas of this section.

5. $\int (3x^4 - x^{-2} + 5) \, dx = \int 3x^4 \, dx -$ _____ $+ \int 5 \, dx$

$= 3$ _____ $- \int x^{-2} \, dx + 5 \int dx$

$= 3(\underline{\hspace{1cm}}) - (\underline{\hspace{1cm}}) + 5(\underline{\hspace{1cm}}) + C$

$=$ _____ .

---

1. $u(x) + C$ 2. $a \int u(x) \, dx$ 3. $\int u(x) \, dx + \int v(x) \, dx$ 4. $\frac{u^{n+1}}{n+1} + C$, -1

5. $\int x^{-2} \, dx$, $\int x^4 \, dx$, $\frac{1}{5} x^5$, $-x^{-1}$, x, $\frac{3}{5} x^5 + x^{-1} + 5x + C$

6. $\int (5x - 3)^9 \, dx.$

    Let $u = 5x - 3$, $du = $ _____ . Thus the integral becomes,

$$\int (5x - 3)^9 \, dx = \tfrac{1}{5} \int \underline{\hspace{1cm}} \, du = \tfrac{1}{5} \left(\underline{\hspace{1.5cm}} \right) + C$$

$$= \tfrac{1}{50} \left(\underline{\hspace{1.5cm}} \right) + C.$$

7. $\int \dfrac{x \, dx}{\left(7 - x^2\right)^5}$.

    Let $u = 7 - x^2$. Then $du = $ _____ so $x \, dx = $ _____ , and

$$\int \dfrac{x \, dx}{\left(7 - x^2\right)^5} = - \tfrac{1}{2} \int \dfrac{du}{\underline{\hspace{0.8cm}}} = - \tfrac{1}{2} \left(\underline{\hspace{1cm}} \right) + C$$

$$= \tfrac{1}{8} \left(\underline{\hspace{1.5cm}} \right) + C .$$

8. $\int \left(x^{3/2} - 2x^{2/3} + 5 \right) \sqrt{x} \, dx = \int \left(x^2 - \underline{\hspace{1.5cm}} + 5x^{1/2} \right) dx$

$$= \left(\underline{\hspace{1cm}} \right) - 2\left(\underline{\hspace{1cm}} \right) + 5\left(\underline{\hspace{1cm}} \right) + C$$

$$= \underline{\hspace{4cm}} .$$

OBJECTIVE B : Solve an elementary differential equation when dy/dx equals an expression in which both x and y may occur, and the variables are separable.

9. The solution technique to solving such an equation is to _____ the variables so that the differential equation is in the form $g(y) \, dy = f(x) \, dx$, and then _____ both sides.

10. To solve the differential equation $\dfrac{dy}{dx} = \dfrac{y^3}{x^2}$ for $x > 0$ and $y > 0$, separate the variables obtaining the equivalent expression $y^{-3} \, dy = $ _____ . Integration of each side then gives $\left(\underline{\hspace{1cm}} \right) = -x^{-1} + C$, or solving for y^2, $y^2 = $ _____ .

11. Let $\dfrac{dy}{dx} = \sqrt{x} + 2x$, $x \geq 0$. Then $dy = \left(\underline{\hspace{1.5cm}} \right) dx$, and integration of both sides gives $y = \left(\underline{\hspace{1.5cm}} \right) + C.$

12. Let $\dfrac{ds}{dt} = \left(t - \dfrac{1}{\sqrt[3]{t}} \right)^2$, $t \neq 0$. Then $ds = \left(t - \dfrac{1}{\sqrt[3]{t}} \right)^2 dt$ or,

    expanding the right side algebraically, $ds = $ _____ . Integration of both sides gives $s = $ _____ .

---

6. $5dx$, u^9, $\dfrac{u^{10}}{10}$, $(5x - 3)^{10}$ 7. $-2x \, dx$, $-\tfrac{1}{2} du$, u^5, $\dfrac{u^{-4}}{-4}$, $\left(7 - x^2\right)^{-4}$

8. $2x^{7/6}$, $\dfrac{x^3}{3}$, $\dfrac{x^{13/6}}{13/6}$, $\dfrac{x^{3/2}}{3/2}$, $\tfrac{1}{3} x^3 - \tfrac{12}{13} x^{13/6} + \tfrac{10}{3} x^{3/2} + C$

9. separate, integrate   10. $x^{-2} \, dx$, $\dfrac{y^{-2}}{-2}$, $x/2(1 - Cx)$

11. $\sqrt{x} + 2x$, $\tfrac{2}{3} x^{3/2} + x^2$ 12. $t^2 - 2t^{2/3} + t^{-2/3}$, $\tfrac{1}{3} t^3 - \tfrac{6}{5} t^{5/3} + 3t^{1/3} + C$

4-2   SELECTING A VALUE FOR THE CONSTANT OF INTEGRATION.

[OBJECTIVE]:   Solve a given differential equation subject to specified initial conditions.

13.   Consider the differential equation $\dfrac{dy}{dx} = \dfrac{x^2 + 1}{\sqrt{y}}$, $y > 0$,

subject to the initial condition $y = 1$ when $x = 0$. Writing the equation in differential form gives, $\sqrt{y}\ dy =$ _____.
Integrating both sides, $\frac{2}{3}\, y^{3/2} =$ _____.   Next we impose the _____ condition to evaluate the constant of integration:
$\frac{2}{3}\, (1)^{3/2} =$ _____;   so $C =$ _____.   Substituting this value of C into the solution of the differential equation gives _____ as the solution to the initial value problem.

14.   Suppose the acceleration of a moving particle is given by the equation $a = \sqrt{t}$,   and that when $t = 0$ it is known that the particle has initial velocity $v = v_0$ and initial position $s = s_0$.   Let us find the equation of motion.   Since $a = dv/dt$, the original equation can be written in differential form as $dv =$ _____.   Integration then gives $v =$ _____ $+\ C_1$.
Imposing the initial condition $v = v_0$ when $t = 0$ yields $v_0 =$ _____;   so $C_1 =$ _____.   Hence the velocity equation becomes $v =$ _____.   Now, $v = ds/dt$ so the last equation can be written in differential form as
$ds = \left(\frac{2}{3}\, t^{3/2} + v_0\right) dt$.   Integration gives $s =$ _____ $+\ C_2$.
From the initial condition $s = s_0$ when $t = 0$ it is readily seen that $C_2 =$ _____.   Therefore, an equation of motion is $s =$ _____ valid for all $t \geq 0$.

4-3   THE SUBSTITUTION METHOD OF INTEGRATION.

[OBJECTIVE]:   Evaluate integals of elementary functions using the substitution method.

15. $\displaystyle\int x\sqrt{3x^2 + 1}\ dx$.   Let $u = 3x^2 + 1$.   Then $du =$ _____   so $x\ dx = \frac{1}{6}\, du$.   Thus the integral becomes

$$\int x\sqrt{3x^2 + 1}\ dx = \int \underline{\hspace{3cm}} = \frac{1}{6} \underline{\hspace{2cm}} + C$$

$$= \underline{\hspace{3cm}}.$$

---

13. $(x^2 + 1)\ dx$, $\frac{1}{3}x^3 + x + C$, initial, $\frac{1}{3}(0)^3 + 0 + C$, $\frac{2}{3}$, $\frac{2}{3}y^{3/2} = \frac{1}{3}x^3 + x + \frac{2}{3}$

14. $\sqrt{t}\ dt$, $\frac{2}{3}t^{3/2}$, $\frac{2}{3}(0)^{3/2} + C_1$, v_0, $\frac{2}{3}t^{3/2} + v_0$, $\frac{4}{15}t^{5/2} + v_0 t$, s_0, $\frac{4}{15}t^{5/2} + v_0 t + s_0$

15. $6x\ dx$, $\frac{1}{6}\sqrt{u}\ du$, $\frac{2}{3}u^{3/2}$, $\frac{1}{9}(3x^2 + 1)^{3/2} + C$

16. $\int \dfrac{1}{z^2 - 6z + 9}\, dz.$ Let $u = z - 3$ so $u^2 =$ _____ and

du = dx. Thus the integral becomes

$$\int \dfrac{1}{z^2 - 6z + 9}\, dz = \int \text{_____}\, du = \text{_____} + C$$

$$= \dfrac{1}{3 - z} + C.$$

4-4 INTEGRALS OF TRIGONOMETRIC FUNCTIONS.

OBJECTIVE : Find indefinite integrals of elementary trigonometric functions. Substitution and algebraic manipulation may be required.

17. $\int \sin (3 - 2x)\, dx.$ Let $u = 3 - 2x.$ Then $du =$ _____ so $dx = -\frac{1}{2}\, du.$ Thus the integral becomes

$$\int \sin (3 - 2x)\, dx = \int \text{_____} = \frac{1}{2} \int \text{_____}$$

$$= \frac{1}{2} \text{_____} + C = \text{_____} .$$

18. $\int x^2 \cos (4x^3)\, dx.$ Let $u = 4x^3.$ Then $du =$ _____ so $x^2\, dx =$ _____. Thus the integral becomes

$$\int x^2 \cos (4x^3)\, dx = \int \cos (4x^3) \cdot x^2\, dx = \int \text{_____}\, du$$

$$= \frac{1}{12} \text{_____} + C = \text{_____} .$$

19. $\int (3 - \sin 2t)^{1/3} \cos 2t\, dt.$ Let $u = 3 - \sin 2t.$ Then $du =$ _____. Substitution into the integral gives

$$\int (3 - \sin 2t)^{1/3} \cos 2t\, dt = \int \text{_____}\, du = \text{_____} + C$$

$$= \text{_____} .$$

20. $\int \sec^{5/2} x \tan x\, dx.$ Let $u = \sec x.$ Then $du =$ _____.

Now, $\sec^{5/2} x \tan x = \sec^{3/2} x \cdot$ _____. Hence substitution into the integral gives,

$$\int \sec^{5/2} x \tan x\, dx = \int \text{_____}\, du = \text{_____} + C$$

$$= \text{_____} .$$

---

16. $z^2 - 6z + 9$, u^{-2}, $-u^{-1}$ 17. $-2dx$, $\sin u \cdot \left(-\frac{1}{2}\right) du$, $-\sin u\, du$, $\cos u$, $\frac{1}{2} \cos (3 - 2x) + C$

18. $12x^2\, dx$, $\frac{1}{12}\, du$, $\frac{1}{12} \cos u$, $\sin u$, $\frac{1}{12} \sin (4x^3) + C$

19. $-2 \cos 2t\, dt$, $-\frac{1}{2} u^{1/3}\, du$, $-\frac{3}{8} u^{4/3}$, $-\frac{3}{8} (3 - \sin 2t)^{4/3} + C$

20. $\sec x \tan x\, dx$, $\sec x \tan x$, $u^{3/2}$, $\frac{2}{5} u^{5/2}$, $\frac{2}{5} \sec^{5/2} x + C$

4-5   DEFINITE INTEGRALS.   THE AREA UNDER A CURVE.

21.   Let $y = f(x)$ define a positive continuous function of   x   on the closed interval $a \leq x \leq b$.   The area under the curve and above the x-axis from $x = a$ to $x = b$ is defined to be the _____ of the sums of the areas of inscribed _____ as their number _____ without bound.

OBJECTIVE A :   Given an elementary function $y = f(x)$ positive and continuous over the interval $a \leq x \leq b$,   approximate the area under the curve by summing a specified number   n   of inscribed rectangles of uniform width.

22.   Consider $y = \cos x$; $a = 0$, $b = \frac{\pi}{2}$, $n = 5$.   The interval $0 \leq x \leq \pi/2$ is divided into _____ subintervals, each of length $\Delta x = ($_____$)/5 = $_____ $\approx .314$ rads.   Thus, the $n - 1 = 4$ intermediate points are located at $x_1 = .314$, $x_2 = .628$, $x_3 = $_____, and $x_4 = $_____.   Now, the curve $y = \cos x$ _____ with   x   over the interval $0 \leq x \leq \pi/2$,   so the altitude of each inscribed rectangle is the length of its _____ edge.   Calculation of the areas of the inscribed rectangles gives (use a calculator or the trigonometric tables for radians in the Appendix of your textbook),

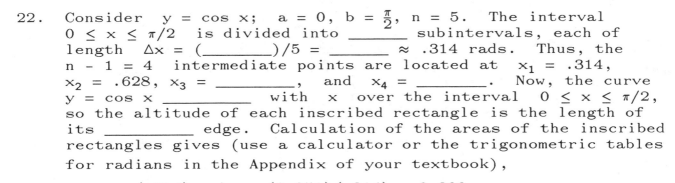

$$\cos (.314) \cdot \Delta x = (0.951)(.314) = 0.299$$

$$\cos (\underline{\quad}) \cdot \Delta x = (0.809)(.314) = \underline{\quad}$$

$$\cos (.942) \cdot \Delta x = (\underline{\quad})(.314) = \underline{\quad}$$

$$\cos (\underline{\quad}) \cdot \Delta x = (\underline{\quad})(.314) = \underline{\quad}$$

$$\cos (\underline{\quad}) \cdot \Delta x = (\underline{\quad})(.314) = \underline{\quad}$$

$$\text{Sum} = \underline{\quad} .$$

The estimate is too _____ because we were using inscribed rectangles.   In fact, our estimate of   0.835   is about   17   percent too small.

OBJECTIVE B :   Interpret and utilize the sigma notation to express or write out sums, and determine (if possible) the value of a sum expressed in sigma notation.

23.   To write out the sum $\sum\limits_{k=3}^{7} 2^{k-2}$,

Replace the   k   in 2^{k-2} by   3   and obtain _____.

Replace the   k   in 2^{k-2} by   4   and obtain _____.

Replace the   k   in 2^{k-2} by   5   and obtain _____.

Replace the   k   in 2^{k-2} by   6   and obtain _____.

---

21.   limit, rectangles, increases

22.   5, $\frac{\pi}{2}$ - 0, $\pi/10$, .942, 1.257, decreases, right, .628, 0.254, 0.588, 0.185, 1.257, 0.309,

0.097, $\pi/2$, 0, 0, 0.835, small

Replace the k in 2^{k-2} by 7 and obtain _____.

The expanded form is $\sum\limits_{k=3}^{7} 2^{k-2}$ = _____.

This finite sum is equal to _____.

24.   To express the finite sum $1 + \frac{1}{2} + \frac{1}{4} + \frac{1}{8} + \frac{1}{16}$ in sigma notation, we may observe that the sum can be written as

$$\left(\frac{1}{2}\right)^0 + \left(\frac{1}{2}\right)^1 + \left(\frac{1}{2}\right)^2 + \left(\underline{\quad}\right)^3 + \left(\underline{\quad}\right)^4 \; .$$

The kth term in this expression is $\left(\frac{1}{2}\right)^k$, and we see that k starts at _____ and ends at _____. Therefore, the required sigma notation is _____.

25.   To write out the sum $\sum\limits_{k=0}^{3} \frac{(-1)^k}{k!} x^k$, first recall that k! means

$1 \cdot 2 \cdot 3 \cdots k$. Then,
Replace the k in $\frac{(-1)^k}{k!}$ by 0 and obtain _____ = _____.

Replace the k in $\frac{(-1)^k}{k!}$ by 1 and obtain _____ = _____.

Replace the k in $\frac{(-1)^k}{k!}$ by 2 and obtain _____ = _____.

Replace the k in $\frac{(-1)^k}{k!}$ by 3 and obtain _____ = _____.

The expanded form is $\sum\limits_{k=0}^{3} \frac{(-1)^k}{k!} x^k$ = _____.

Substitution of x = 1 gives the value ___.

26.   The limit, $\lim \sum f(c_k) \Delta_k x$, is called the _____ of f from a to b. It is a number. This number is denoted by the symbol _____. The number a in the symbol is called the _____ _____ of integration, and _____ is the upper limit.

27.   For k any constant, $\int_a^b kf(x)\ dx$ = _____.

28. $\int_a^b [f(x) + g(x)]\ dx$ = _____.

29. $\int_b^a f(x)\ dx$ = - _____.

---

23. 2^1, 2^2, 2^3, 2^4, 2^5, $2^1 + 2^2 + 2^3 + 2^4 + 2^5$, 62   24. $\frac{1}{2}$, $\frac{1}{2}$, 0, 4, $\sum\limits_{k=0}^{4} \left(\frac{1}{2}\right)^k$

25. $\frac{(-1)^0}{0!} x^0 = 1$, $\frac{(-1)^1}{1!} x^1 = -x$, $\frac{(-1)^2}{2!} x^2 = \frac{1}{2} x^2$, $\frac{(-1)^3}{3!} x^3 = -\frac{1}{6} x^3$, $1 - x + \frac{1}{2} x^2 - \frac{1}{6} x^3$, $\frac{1}{3}$

26. definite integral, $\int_a^b f(x)$, lower limit, b   27. $k \int_a^b f(x)\ dx$

28. $\int_a^b f(x)\ dx + \int_a^b g(x)\ dx$ 29. $\int_a^b f(x)\ dx$

30.   If f is positive-valued and continuous over $a \leq x \leq b$, then $\int_a^b f(x) \, dx$ is the _____ under the graph of $y = f(x)$, above the _____ from $x = a$ to $x = b$.

31.   If f is both positive-valued and negative-valued over $a \leq x \leq b$, then $\int_a^b f(x) \, dx$ is the algebraic sum of the _____ areas bounded by the graph $y = f(x)$ and the x-axis, from $x = a$ to $x = b$.

32.   If $f(x) \geq 0$ for $a \leq x \leq b$, then $\int_a^b f(x) \, dx$ is _____.

33.   If $f(x) < g(x)$ for $a \leq x \leq b$, then their definite integrals are related by _____.

4-6   CALCULATING DEFINITE INTEGRALS BY SUMMATION.

OBJECTIVE A: Establish elementary arithmetic formulas by the method of mathematical induction. That is, show the formula is correct for $n = 1$; and show that if true for n, the formula is also true for $n + 1$.

34.   Consider the formula $2^1 + 2^2 + 2^3 + \ldots + 2^n = 2^{n+1} - 2$. In sigma notation the formula could be written as $\sum_{k=1}^{n}$ _____ = _____.   To verify the formula for $n = 1$:

$$\sum_{k=1}^{1} 2^k = 2^{1+1} - 2 \quad \text{or} \quad 2^1 = 2^{1+1} - 2.$$

Is this last statement true?

35.   Write the formula for $n = 2$. Is it true? (It is not necessary to verify the formula for $n = 2$ when applying mathemtical induction, but this problem gives you some practice in working with the formula.)

36.   Suppose now that n is any integer for which the formula in Problem 34 is known to be true (at the moment, n could be the integers ___ and ____). Let us add 2^{n+1} to both sides of the formula: $2^1 + 2^2 + 2^3 + \ldots + 2^n + 2^{n+1} = (2^{n+1} - 2) + 2^{n+1}$ which must also be true for the same n. The right side of this last equation can be written as $(2^{n+1} - 2) + 2^{n+1} = 2(\underline{\quad}) - 2 = \underline{\qquad} - 2$. Thus, $2^1 + 2^2 + \ldots + 2^n + 2^{n+1} = \underline{\qquad}$. This is just like the original formula in Problem 34 with n replaced by _____.

---

30.   area, x-axis   31.   signed   32.   nonnegative or ≥ 0 33. $\int_a^b f(x) \, dx \leq \int_a^b g(x) \, dx$

34. $\sum_{k=1}^{n} 2^k = 2^{n+1} - 2$, yes because $2 = 4 - 2$ 35. $2^1 + 2^2 = 2^{2+1} - 2$, yes because $2 + 4 = 8 - 2$

36.   1, 2, 2^{n+1}, 2^{n+2}, $2^{n+2} - 2$ or $2^{(n+1)+1} - 2$, $n + 1$

OBJECTIVE B : Find the area bounded by a curve $y = f(x)$ positive over $a \le x \le b$ by finding the limit of the sum of inscribed or circumscribed rectangles, if f is of the form $y = mx^k$ for $k = 0,1,2,3$.

37. To find the area under the graph $y = x^2$, $-1 \le x \le 0$ using inscribed rectangles, divide the interval $-1 \le x \le 0$ into n subintervals each of equal length $\Delta x = $ _____ by inserting the points $x_1 = -1 + \Delta x$, $x_2 = -1 + 2\Delta x$, ..., $x_{n-1} = $ _____. Notice that $y = x^2$ is a _____ function of x over the interval $-1 \le x \le 0$. Thus the height of the first inscribed rectangle is _____. The inscribed rectangles have areas

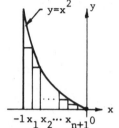

$$f(x_1)\Delta x = (-1 + \Delta x)^2 \cdot \Delta x$$
$$f(x_2)\Delta x = (-1 + 2\Delta x)^2 \cdot \Delta x$$
.
.
.
$$f(x_{n-1})\Delta x = (-1 + (n-1)\Delta x)^2 \cdot \Delta x$$
$$f(\underline{\quad})\Delta x = \underline{\hspace{3cm}},$$

whose sum is
$$S_n = \left[\left(-1 + \tfrac{1}{n}\right)^2 + \left(-1 + \tfrac{2}{n}\right)^2 + \ldots + \left(-1 + \tfrac{n-1}{n}\right)^2 + 0\right] \cdot \underline{\hspace{2cm}}$$

Expanding each term on the right side and collecting like terms gives,

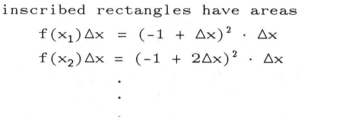

$$S_n = \left((-1)^2(n - 1) + 2(-1)\left(\tfrac{1}{n} + \tfrac{2}{n} + \ldots + \tfrac{n-1}{n}\right)\right.$$
$$\left. + \left(\tfrac{1^2}{n^2} + \tfrac{2^2}{n^2} + \ldots + \tfrac{(n-1)^2}{n^2}\right)\right) \tfrac{1}{n}$$
$$= \tfrac{1}{n}\left[(n - 1) - \tfrac{2}{n}\left(\underline{\hspace{2cm}}\right) + \tfrac{1}{n^2}\left(\underline{\hspace{2cm}}\right)\right]$$
$$= \tfrac{1}{n}\left[(n - 1) - \tfrac{2}{n} \cdot \left(\underline{\hspace{1.5cm}}\right) + \tfrac{1}{n^2} \cdot \left(\underline{\hspace{2cm}}\right)\right]$$
$$= \tfrac{1}{n}\left[(n - 1) - \underline{\hspace{1cm}} + \tfrac{1}{6}\left(2n - 3 + \tfrac{1}{n}\right)\right] = \underline{\hspace{3cm}}.$$

The area under the graph is defined to be the limit of _____ as $n \to \infty$. This limit equals _____.

38. To find the area under the graph $y = 3x^2$, $3 \le x \le 4$ using circumscribed rectangles, divide the interval $3 \le x \le 4$ into n subintervals each of equal length $\Delta x = $ _____ by inserting the points
$$x_1 = 3 + \Delta x, \quad x_2 = 3 + 2\Delta x, \quad \ldots, \quad x_{n-1} = \underline{\hspace{2cm}}.$$

---

37. $1/n$, $-1 + (n - 1)\Delta x$, decreasing, $(-1 + \Delta x)^2$ or x_1^2, 0, $0 \cdot \Delta x$ or $(-1 + n\Delta x)^2 \cdot \Delta x$, $1/n$,

$1 + 2 + 3 + \ldots + (n - 1)$, $1^2 + 2^2 + \ldots + (n - 1)^2$, $\dfrac{(n - 1)n}{2}$, $\dfrac{(n - 1)n\left(2(n - 1) + 1\right)}{6}$, $(n - 1)$,

$\dfrac{1}{3} - \dfrac{1}{2n} + \dfrac{1}{6n^2}$, S_n, $1/3$

Notice that $y = 3x^2$ is an _____ function of x over the interval $3 \le x \le 4$. Thus, the height of the first circumscribed rectangle is _____; the height of the last circumscribed rectangle is _____. The circumscribed rectangles have areas

$$f(x_1)\Delta x = 3(3 + \Delta x)^2 \cdot \Delta x$$

$$f(x_2)\Delta x = 3(3 + 2\Delta x)^2 \cdot \Delta x$$

$$\cdot$$
$$\cdot$$
$$\cdot$$

$$f(\underline{\quad})\Delta x = \underline{\qquad\qquad} ,$$

whose sum is

$$S_n = \left[3\left(\underline{\quad}\right)^2 + 3\left(3 + \tfrac{2}{n}\right)^2 + \cdots + 3\left(3 + \tfrac{n}{n}\right)^2 \right]\tfrac{1}{n}$$

$$= \tfrac{3}{n}\left[n3^2 + 6\left(\underline{\qquad\qquad}\right) + \left(\underline{\qquad\qquad}\right) \right]$$

$$= \tfrac{3}{n}\left[n3^2 + \tfrac{6}{n}\cdot\left(\underline{\qquad\qquad}\right) + \tfrac{1}{n^2}\cdot\left(\underline{\qquad\qquad}\right) \right]$$

$$= 3\left[3^2 + 3\left(1 + \tfrac{1}{n}\right) + \tfrac{1}{6}\left(\underline{\qquad\qquad}\right) \right].$$

Hence, $\lim S_n$ as $n \to \infty$ equals _____ $= 37$, the area.

4-7   THE FUNDAMENTAL THEOREMS OF INTEGRAL CALCULUS.

39.   The First Fundamental Theorem of Calculus concerns the integral
$$F(x) = \int_a^x f(t)\,dt \ ,$$

where f is continuous on $[a,b]$. This theorem says that F is _____ at every point x in $[a,b]$ and
$$F'(x) = \underline{\qquad\qquad}.$$

40.   The notation $g(x)]_d^c$ means _____.

41.   The Second Fundamental Theorem of Calculus gives a rule for calculating the definite integral $\int_a^b f(x)\,dx$ of a continuous function. The rule states that you must first find an _____ F of f. That is, the relationship between F and f is _____. Next, calculate the number $F(b) - $ _____. This computation gives
$$\int_a^b f(x)\,dx = \underline{\qquad\qquad}.$$

---

38. $1/n$, $3 + (n - 1)\Delta x$, increasing, $3x_1{}^2$ or $3(3 + \Delta x)^2$, $3(4)^2$ or $3(3 + n\Delta x)^2$, 4, $3(4)^2 \cdot \Delta x$

or $3(3 + n\Delta x)^2 \cdot \Delta x$, $3 + \tfrac{1}{n}$, $\tfrac{1}{n} + \tfrac{2}{n} + \cdots + \tfrac{n}{n}$, $\tfrac{1^2}{n^2} + \tfrac{2^2}{n^2} + \cdots + \tfrac{n^2}{n^2}$, $\tfrac{n(n + 1)}{2}$,

$\tfrac{n(n + 1)(2n + 1)}{6}$, $2 + \tfrac{3}{n} + \tfrac{1}{n^2}$, $3(3^2 + 3 + \tfrac{1}{3})$ 39. differentiable, $f(x)$ 40. $g(c) - g(d)$

41. antiderivative, $F'(x) = f(x)$, $F(a)$, $F(b) - F(a)$

$\boxed{\text{OBJECTIVE A}}$:   Evaluate definite integrals of elementary continuous functions, using the Integral Evaluation Theorem (Second Fundamental Theorem).

42.   Find $\int_{-1}^{2} (x^3 - 2x + 5)\ dx$.

$$\int_{-1}^{2} (x^3 - 2x + 5)\ dx = \underline{\hspace{3cm}} \Big]_{-1}^{2}$$
$$= \left(\tfrac{1}{4} (2^4) - 2^2 + 5\cdot 2\right) - \left(\underline{\hspace{3cm}}\right)$$
$$= 10 - (\underline{\hspace{1.5cm}}) = \underline{\hspace{1.5cm}}.$$

43.   Find $\int_{0}^{\pi/4} (3 - \sin 2t)^{1/3} \cos 2t\ dt$.

Solution.   From Problem 19 of this chapter, the indefinite integral $\int (3 - \sin 2t)^{1/3} \cos 2t\ dt = \underline{\hspace{4cm}}.$
Therefore, the definite integral is given by

$$\int_{0}^{\pi/4} (3 - \sin 2t)^{1/3} \cos 2t\ dt = \underline{\hspace{4cm}} \Big]_{0}^{\pi/4}$$
$$= -\tfrac{3}{8} (\underline{\hspace{1.5cm}})^{4/3} + \tfrac{3}{8} (\underline{\hspace{1.5cm}})^{4/3}$$
$$= \underline{\hspace{3cm}} \approx 0.67759.$$

$\boxed{\text{OBJECTIVE B}}$:   Find the area under the graph of a positive-valued continuous function $y = f(x)$ over an interval $a \le x \le b$ by integration.

44.   To find the area under the graph of $y = \sqrt{9 + x}$ for $-9 \le x \le 0$, we first find an $\underline{\hspace{2cm}}$ integral for $y = f(x)$:
$$\int \sqrt{9 + x}\ dx = \underline{\hspace{3cm}} + C.$$
We then compute the area: $\underline{\hspace{3cm}} \Big]_{-9}^{0} = \underline{\hspace{2.5cm}}.$
Notice that the constant C of integration does not enter into the final calculation.

45.   Find the area between the curve $y = x^2 + x + 1$, the x-axis, and the lines $x = 2$, $x = 3$.

Solution.   The area is
$$A_2^3 = \int (x^2 + x + 1)\ dx\Big]_2^3 = \underline{\hspace{3cm}}\Big]$$
$$= \left(\tfrac{27}{3} + \underline{\hspace{1cm}} + 3\right) - \left(\underline{\hspace{1cm}} + \tfrac{4}{2} + 2\right) = \underline{\hspace{1.5cm}}.$$

46.   To find $\int_{0}^{1} \dfrac{x\ dx}{\sqrt{4 - x^2}}$, first find $\int \dfrac{x\ dx}{\sqrt{4 - x^2}}$.   Using substitution, let $u = 4 - x^2$.   Then $du = \underline{\hspace{1.5cm}}$ so $x\ dx = \underline{\hspace{1cm}}.$
Substitution into the indefinite integral yields,

---

42. $\tfrac{1}{4} x^4 - x^2 + 5x,\ \ \tfrac{1}{4} (-1)^4 - (-1)^2 + 5(-1),\ \ -\tfrac{23}{4},\ \tfrac{63}{4},$

43.· $-\tfrac{3}{8} (3 - \sin 2t)^{4/3} + C,\ \ -\tfrac{3}{8} (3 - \sin 2t)^{4/3},\ \ 3 - 1,\ \ 3 - 0,\ \ \tfrac{3}{8} (3^{4/3} - 2^{4/3})$

44.   indefinite, $\tfrac{2}{3} (9 + x)^{3/2},\ \ \tfrac{2}{3} (9 + x)^{3/2},\ \ 18$ 45. $\tfrac{1}{3} x^3 + \tfrac{1}{2} x^2 + x\big]_2^3,\ \tfrac{9}{2},\ \tfrac{8}{3},\ \tfrac{59}{6}$

$$\int \frac{x \, dx}{\sqrt{4 - x^2}} = \underline{\hspace{2cm}} = -\frac{1}{2} \left(\underline{\hspace{1.5cm}} \right) + C = -\sqrt{4 - x^2} + C.$$

Thus, $\displaystyle\int_0^1 \frac{x \, dx}{\sqrt{4 - x^2}} = \underline{\hspace{2cm}} \Big]_0^1 = \underline{\hspace{2cm}} \approx 0.268.$

OBJECTIVE C : Use the Second Fundamental Theorem to calculate the derivative of an integral $\displaystyle\int_0^{v(x)} f(t) \, dt$ with respect to x. Assume that the integrand f is continuous and that v is a differentiable function of x.

47. If $F(x) = \displaystyle\int_1^x (t^5 - 2t^3 + 1)^4 \, dt,$ then we may find $F'(x)$ by replacing t by x in the integrand. Thus, $F'(x) = \underline{\hspace{2cm}}.$

48. Let $F(x) = \displaystyle\int_0^{x^2} \sqrt{1 + t} \, dt.$ If $u = x^2,$ then by the chain rule,
$\dfrac{dF}{dx} = \dfrac{dF}{du} \cdot \underline{\hspace{1cm}}.$ Now, $\dfrac{dF}{du} = \dfrac{d}{du} \displaystyle\int_0^u \sqrt{1 + t} \, dt = \underline{\hspace{2cm}}.$ Thus,
$F'(x) = \sqrt{1 + u} \cdot \underline{\hspace{1cm}} = \underline{\hspace{2cm}}.$

4-8   SUBSTITUTION IN DEFINITE INTEGRALS.

OBJECTIVE : Apply the Substitution Formula for Definite Integrals to evaluate them.

49. $\displaystyle\int_0^{\pi/6} \frac{\cos x \, dx}{\sqrt{1 - \sin x}}.$ Let $u = \sin x.$ Then $du = \underline{\hspace{2cm}}.$
Thus,
$$\int_0^{\pi/6} \frac{\cos x \, dx}{\sqrt{1 - \sin x}} = \int_{\underline{\hspace{0.3cm}}}^{\underline{\hspace{0.3cm}}} \frac{du}{\sqrt{1 - u}} = \underline{\hspace{2cm}} = -2\left(\underline{\hspace{1cm}} \right) = 2 - \sqrt{2}.$$

50. $\displaystyle\int_0^1 \frac{x \, dx}{\sqrt{4 - x^2}}.$ Let $u = 4 - x^2.$ Then,
$$\int_0^1 \frac{x \, dx}{\sqrt{4 - x^2}} = -\frac{1}{2} \int_{\underline{\hspace{0.3cm}}}^{\underline{\hspace{0.3cm}}} \frac{du}{\sqrt{u}} = \underline{\hspace{2cm}} = -\sqrt{3} + 2$$
in agreement with the result obtained in Problem 46.

---

46. $-2x\,dx,$ $-\frac{1}{2} \, du,$ $-\frac{1}{2} \displaystyle\int \frac{du}{u^{1/2}},$ $2u^{1/2},$ $-\sqrt{4 - x^2},$ $-\sqrt{3} + 2$

47. $(x^5 - 2x^3 + 1)^4$

48. $\dfrac{du}{dx},$ $\sqrt{1 + u},$ $2x,$ $2x\sqrt{1 + x^2}$

49. $\cos x \, dx,$ $\displaystyle\int_0^{1/2},$ $-2\sqrt{1 - u}\Big]_0^{1/2},$ $\dfrac{1}{\sqrt{2}} - 1$

50. $\displaystyle\int_4^3,$ $-\sqrt{u}\Big]_4^3$

4-9   RULES FOR APPROXIMATING DEFINITE INTEGRALS.

OBJECTIVE A : Approximate a given definite integral by use of the
trapezoidal rule for a specified number n of
subdivisions. Estimate the error in this
approximation.

51.   Let $y = f(x)$ be defined and continuous over the interval
$a \leq x \leq b$. Divide the interval $[a,b]$ into n subintervals,
each of length $h = (b - a)/n$, by inserting the points
$x_1 = a + h$, $x_2 = a + 2h$, . . ., $x_{n-1} = a + (n - 1)h$. Set
$x_0 = a$ and $x_n = b$ for convenience in notation. Define
$y_k = f(x_k)$ for each $k = 0,1,2,...,n$. Then the trapezoidal
approximation for the definite integral is

$$\int_a^b f(x) \; dx = \underline{\hspace{5cm}} .$$

The error estimate is $E = \underline{\hspace{3cm}}$, for some number c
satisfying $\underline{\hspace{3cm}}$.

52.   Use the trapezoidal rule to approximate $\int_0^1 \sqrt{1 + x^2} \; dx$, $n = 5$.
Solution. Here, $h = \dfrac{1 - 0}{5} = \underline{\hspace{1.5cm}}$, $x_0 = \underline{\hspace{1.5cm}}$, $x_5 = \underline{\hspace{1.5cm}}$.
The subdivision points are $x_1 = \frac{1}{5}$, $x_2 = \underline{\hspace{1.5cm}}$, $x_3 = \underline{\hspace{1.5cm}}$,
$x_4 = \underline{\hspace{1.5cm}}$. The corresponding function values are computed as,
$y_0 = \sqrt{1 + 0^2} = 1$, $y_1 = \sqrt{1 + \left(\frac{1}{5}\right)^2} = \underline{\hspace{1.5cm}} \approx \underline{\hspace{2cm}}$
$y_2 = \sqrt{1 + \left(\frac{2}{5}\right)^2} = \underline{\hspace{1.5cm}} \approx \underline{\hspace{2.5cm}}$, $y_3 \approx \underline{\hspace{2.5cm}}$,
$y_4 \approx \underline{\hspace{2cm}}$, and $y_5 \approx \underline{\hspace{2cm}}$. Therefore, the
trapezoidal approximation is

$T = \dfrac{1}{2(5)} \cdot (y_0 + 2y_1 + 2y_2 + 2y_3 + 2y_4 + y_5)$,   or

$T = \dfrac{1}{10} \cdot (1 + 2.03960 + \underline{\hspace{1.5cm}} + \underline{\hspace{1.5cm}} + \underline{\hspace{1.5cm}} + \underline{\hspace{1.5cm}})$

$\;\; = \dfrac{1}{10} \cdot (\underline{\hspace{2cm}}) = \underline{\hspace{2.5cm}}$.

Therefore,

$$\int_0^1 \sqrt{1 + x^2} \; dx \approx \underline{\hspace{2.5cm}} .$$

53.   To estimate the error in the approximation of Problem 52, let
$f(x) = \sqrt{1 + x^2}$. Then, $f'(x) = \underline{\hspace{2cm}}$ and $f''(x) = \underline{\hspace{2.5cm}}$.
Therefore, for $0 \leq x \leq 1$, we see that

$$|f''(x)| = \frac{1}{\left(1 + x^2\right)^{3/2}} < \underline{\hspace{2cm}} .$$

Thus, the error E satisfies

$$E = \left| \frac{b-a}{12} f''(c) \cdot h^2 \right| < \left| \underline{\hspace{1cm}} \cdot \underline{\hspace{1cm}} \cdot \underline{\hspace{1cm}} \right| \approx \underline{\hspace{2.5cm}} .$$

---

51. $\frac{h}{2} (y_0 + 2y_1 + 2y_2 + ... + 2y_{n-1} + y_n)$, $\frac{b-a}{12} f''(c) \cdot h^2$, $a < c < b$

52. $\frac{1}{5}$, 0, 1, $\frac{2}{5}$, $\frac{3}{5}$, $\frac{4}{5}$, $\frac{\sqrt{26}}{5}$, 1.01980, $\frac{\sqrt{29}}{5}$, 1.07703, 1.16619, 1.28062, 1.41421, 2.15407,

2.33238, 2.56124, 1.41421, 11.50150, 1.15015, 1.15015

Then, $\int_0^1 \sqrt{1 + x^2}\ dx = 1.15015 \pm E$,   or

$$\underline{\hspace{3cm}} < \int_0^1 \sqrt{1 + x^2}\ dx < \underline{\hspace{3cm}}.$$

(In fact, the actual value of the integral to five decimal places is 1.14779. Thus the error is about 2/10 of one percent.)

54. How many subdivisions are required to obtain $\int_0^1 \sqrt{1 + x^2}\ dx$ to 5 decimal places of accuracy by the trapezoidal rule?

Solution. From Problem 53, $E = \left|\frac{1}{12}\ f''(c)\ h^2\right| < \frac{1}{12}\ h^2$. In order to obtain 5 place accuracy, we need $E < 5 \cdot 10^{-6}$. Thus, $\frac{1}{12}\ h^2 < 5 \cdot 10^{-6}$ implies $h^2 < \underline{\hspace{2cm}}$ or, since $h = \frac{b - a}{n}$

$= \underline{\hspace{2cm}}$, $n^2 > \underline{\hspace{2cm}}$. Thus, $n > \underline{\hspace{1.5cm}} \approx \underline{\hspace{1cm}}$, and choosing $n = \underline{\hspace{1cm}}$ as the number of subdivisions ensures 5 place accuracy. (This is only an upper estimate: fewer subdivisions may work, but there are no guarantees.)

OBJECTIVE B : Approximate a given definite integral by use of Simpson's rule for a specified even number n of subdivisions. Estimate the error in this approximation.

55. Let $y = f(x)$ be defined and continuous over the interval $a \le x \le b$. Divide the interval $[a,b]$ into n subintervals, where n is an <u>even</u> number, each of length $h = (b - a)/n$, using the points $x_0 = a$, $x_1 = a + h$, $x_2 = 1 + 2h$, ..., $x_n = a + nh = b$. Define $y = f(x_k)$ for each $k = 0,1,2,\ldots,n$. Then the Simpson approximation for the definite integral is

$$\int_a^b f(x)\ dx \underline{\hspace{6cm}}.$$

The error estimate is $E = \underline{\hspace{4cm}}$.

56. Simpson's rule is exact if $f(x)$ is $\underline{\hspace{4cm}}$.

57. Approximate $\int_0^1 \sqrt{1 + x^2}\ dx$ by Simpson's rule with $n = 6$.
Solution. Here $h = \underline{\hspace{1cm}}$, $x_0 = \underline{\hspace{1cm}}$, and $x_6 = \underline{\hspace{0.7cm}}$. The subdivision points are,

$x_1 = \frac{1}{6}$, $x_2 = \underline{\hspace{1.5cm}}$, $x_3 = \underline{\hspace{1.5cm}}$, $x_4 = \underline{\hspace{1.5cm}}$, and $x_5 = \underline{\hspace{1.5cm}}$.

---

53. $x(1 + x^2)^{-1/2}$, $(1 + x^2)^{-3/2}$, 1, $\left|\frac{1}{12} \cdot 1 \cdot \frac{1}{25}\right|$, 0.00333, 1.14682, 1.15348

54. $60 \cdot 10^{-6}$, $\frac{1}{n}$, $\frac{1}{60} \cdot 10^6$, $\frac{1}{2\sqrt{15}}\ 10^3$, 129, 130

55. $\frac{h}{3}(y_0 + 4y_1 + 2y_2 + 4y_3 + \ldots + 2y_{n-2} + 4y_{n-1} + y_n)$, $\frac{b-a}{180}\ f^{(4)}(c) \cdot h^4$

56. a polynomial of degree < 4

The corresponding function values are $y_0 = 1$,
$y_1 = \sqrt{1 + \left(\frac{1}{6}\right)^2} =$ _____ $\approx$ _____ , $y_2 \approx$ _____ ,

$y_3 \approx$ _____ , $y_4 \approx$ _____ , $y_5 \approx$ _____ , and

$y_6 =$ _____ $\approx$ _____ . Therefore, the Simpson approximation is,

$$S = \frac{1}{6 \cdot 3} (y_0 + 4y_1 + 2y_2 + 4y_3 + 2y_4 + 4y_5 + y_6)$$

$$= \frac{1}{18} (1 + 4.05518 + \underline{\qquad} + \underline{\qquad} + \underline{\qquad} + \underline{\qquad} + \underline{\qquad})$$

$$= \frac{1}{18} (\underline{\qquad}) = \underline{\qquad} .$$

Compare the answer with that found in Problem 52 where we employed the trapezoidal rule.

58. To estimate the error in the approximation in Problem 57, let $f(x) = \sqrt{1 + x^2}$. Then, from Problem 53,
$f''(x) = \left(1 + x^2\right)^{-3/2}$, so that $f^{(3)}(x) =$ _____ and
$f^{(4)}(x) =$ _____ . Now,

$$\frac{4x^2 - 1}{\left(1 + x^2\right)^{7/2}} < \frac{4x^2}{\left(1 + x^2\right)^{7/2}} < \frac{4x^2}{1 + x^2} < 4 \quad \text{for} \quad 0 \le x \le 1.$$

Thus, the error E satisfies

$$E = \left| \frac{b - a}{180} f^{(4)}(c) \cdot h^4 \right| < |\underline{\qquad} \cdot \underline{\qquad} \cdot \underline{\qquad}| \approx \underline{\qquad} .$$

Therefore, we observe that Simpson's rule, with $n = 6$, provides the value of the integral to at least 3 decimal places of accuracy; with $n = 10$ we will obtain at least 4 place accuracy, according to our error estimates.

---

57. $\frac{1}{6}$, 0, 1, $\frac{1}{3}$, $\frac{1}{2}$, $\frac{2}{3}$, $\frac{5}{6}$, $\frac{\sqrt{37}}{6}$, 1.01379, 1.05409, 1.11803, 1.20185, 1.30171, $\sqrt{2}$, 1.41421,

2.10819, 4.47212, 2.40370, 5.20684, 1.41421, 20.66021, 1.14779

58. $-3x\left(1 + x^2\right)^{-5/2}$, $3\left(4x^2 - 1\right)\left(1 + x^2\right)^{-7/2}$, $\frac{1}{180} \cdot 12 \cdot \frac{1}{6^4}$, 0.00005

CHAPTER 4   SELF-TEST

1. Find the following indefinite integrals.

    (a) $\displaystyle\int (x - 1)(2 + x)\ dx$

    (b) $\displaystyle\int \sqrt{2x - 1}\ dx$

    (c) $\displaystyle\int x^2(5 - 3x^3)^{-1/2}\ dx$

    (d) $\displaystyle\int x^{-1/2} \sin(\sqrt{x} - 3)\ dx$

2. Solve the differential equations.

    (a) $\dfrac{dy}{dx} = 4x - 3$

    (b) $\dfrac{dy}{dx} = y^{-2} \sec^2 x;\quad y = 3\quad$ if $\quad x = 0$

3. Approximate the area under the curve $y = x^2 - 2x + 4$ between $x = 1$ and $x = 4$ by summing $n = 6$ inscribed rectangles of uniform width.

4. Find the numerical values of each of the following.

    (a) $\displaystyle\sum_{k=1}^{4} \dfrac{1}{2k}$

    (b) $\displaystyle\sum_{n=1}^{5} n(n - 3)$

    (c) $\displaystyle\sum_{k=5}^{6} (2k - 1)$

5. Establish the formula $1 + 4 + 7 + \ldots + (3n - 2) = \dfrac{n(3n - 1)}{2}$ by mathematical induction.

6. Find the area under the curve $y = 1 - 2x$ over the interval $-2 \le x \le 0$ by taking the limit of the sum of circumscribed rectangles.

7. Find the area under the curve $y = x\sqrt{x^2 + 1}$, above the x-axis, between $x = 1$ and $x = 4$.

8. Evaluate the definite integrals.

    (a) $\displaystyle\int_{1}^{4} \dfrac{(x - 2)^2}{\sqrt{x}}\ dx$

    (b) $\displaystyle\int_{-2}^{0} x^2(4 - x)\ dx$

9. Find $\dfrac{d}{dx} \displaystyle\int_{1-x^2}^{0} \sqrt[3]{t^2 + 1}\ dt$.

10. Use the trapezoidal rule to approximate $\displaystyle\int_{0}^{1} \sqrt{1 + x^3}\ dx$ with $n = 4$.

11. Use Simpson's rule to approximate $\displaystyle\int_{1}^{2} \dfrac{dx}{x}$ with $n = 6$. Estimate the error in your approximation.

12. Compute $\displaystyle\int_{0}^{\pi} f(x)\ dx$ where $f(x) = \begin{cases} \sin x, & 0 \le x < \frac{\pi}{2} \\ \pi x, & \frac{\pi}{2} \le x \le \pi. \end{cases}$

SOLUTIONS TO CHAPTER 4 SELF-TEST

1. (a) $\int (x - 1)(2 + x)\ dx = \int (x^2 + x - 2)\ dx = \frac{1}{3}x^3 + \frac{1}{2}x^2 - 2x + C$

   (b) $\int \sqrt{2x - 1}\ dx = \frac{1}{2} \int \sqrt{u}\ du \quad (u = 2x - 1) = \frac{1}{3}(2x - 1)^{3/2} + C$

   (c) $\int x^2(5 - 3x^3)^{-1/2}\ dx = -\frac{1}{9} \int u^{-1/2}\ du \quad (u = 5 - 3x^3)$

   $\qquad\qquad\qquad = -\frac{2}{9}(5 - 3x^3)^{1/2} + C$

   (d) $\int x^{-1/2} \sin (\sqrt{x} - 3)\ dx = 2 \int \sin u\ du \quad (u = \sqrt{x} - 3)$

   $\qquad\qquad\qquad\qquad = -2 \cos (\sqrt{x} - 3) + C$

2. (a) $dy = (4x - 3)\ dx \quad$ or $\quad y = \int (4x - 3)\ dx = 2x^2 - 3x + C$

   (b) $y^2\ dy = \sec^2 x\ dx, \quad$ so $\quad \int y^2\ dy = \int \sec^2 x\ dx \quad$ and then
   $\frac{1}{3} y^3 = \tan x + C. \quad$ Since $\quad y = 3 \quad$ if $\quad x = 0,$

   $\frac{1}{3}(3)^3 = \tan 0 + C \quad$ or $\quad C = 9. \quad$ Hence, the solution is

   $\frac{1}{3} y^3 = \tan x + 9.$

3. The partition points are $x_0 = 1, \quad x_1 = 3/2, \quad x_2 = 2, \quad x_3 = 5/2,$
   $x_4 = 3, \quad x_5 = 7/2, \quad$ and $\quad x_6 = 4.$ Since $dy/dx > 0$ on the
   interval $1 \leq x \leq 4,$ the curve is increasing so the altitude of
   each rectangle is its left edge. Thus, the areas of the
   inscribed rectangles for $y = f(x)$ are,

   $f(1)\ \Delta x = 3 \cdot \frac{1}{2} = \frac{3}{2}$
   $f(\frac{3}{2})\ \Delta x = \frac{13}{4} \cdot \frac{1}{2} = \frac{13}{8}$
   $f(2)\ \Delta x = 4 \cdot \frac{1}{2} = 2$
   $f(\frac{5}{2})\ \Delta x = \frac{21}{4} \cdot \frac{1}{2} \cdot \frac{21}{8}$
   $f(3)\ \Delta x = 7 \cdot \frac{1}{2} = \frac{7}{2}$
   $f(\frac{7}{2})\ \Delta x = \frac{37}{4} \cdot \frac{1}{2} = \frac{37}{8}$

   $\qquad$ Sum $= \frac{127}{8} = 15.875, \quad$ approximate area

4. (a) $\frac{1}{2} + \frac{1}{4} + \frac{1}{6} + \frac{1}{8} = \frac{25}{24}$

   (b) $1(-2) + 2(-1) + 3(0) + 4(1) + 5(2) = 10$

   (c) $(2 \cdot 5 - 1) + (2 \cdot 6 - 1) = 9 + 11 = 20$

5. For $n = 1, \quad (3 \cdot 1 - 2) = \dfrac{1(3 \cdot 1 - 1)}{2} \quad$ or, $\quad 1 = \frac{2}{2} \quad$ is true.
   Suppose the formula is true for the known integer $n.$ Adding
   $3(n + 1) - 2 = 3n + 1 \quad$ to both sides of the formula gives,

$$1 + 4 + 7 + \ldots + (3n-2) + (3n+1) = \frac{n(3n-1)}{2} + (3n+1)$$

$$= \frac{3n^2 - n + 2(3n + 1)}{2}$$

$$= \frac{3n^2 + 5n + 2}{2} = \frac{(n+1)(3n+2)}{2},$$

which is exactly like the original formula if we replace n by n + 1.

6. Subdivision of the interval $-2 \leq x \leq 0$ into n equal pieces each of length $\Delta x = \frac{0 - (-2)}{n} = \frac{2}{n}$ gives the subdivision points $x_0 = -2$, $x_1 = -2 + \frac{2}{n}$, $x_2 = -2 + \frac{4}{n}$, ..., $x_{n-1} = -2 + \frac{2(n - 1)}{n}$, $x_n = 0$. Since $y = 1 - 2x$ is decreasing over the interval, we take the left edge for the altitude of each circumscribed rectangle. These rectangles have area

$$f(-2) \ \Delta x = (1 + 4) \cdot \frac{2}{n} = 5 \cdot \frac{2}{n}$$

$$f(-2 + \tfrac{2}{n}) \ \Delta x = (1 + 4 - \tfrac{4}{n}) \cdot \tfrac{2}{n} = (5 - \tfrac{4}{n}) \cdot \tfrac{2}{n}$$

$$f(-2 + \tfrac{4}{n}) \ \Delta x = (1 + 4 - \tfrac{8}{n}) \cdot \tfrac{2}{n} = (5 - \tfrac{8}{n}) \cdot \tfrac{2}{n}$$

$$\vdots$$

$$f(-2 + \tfrac{2n - 2}{n}) \Delta x = \left[1 + 4 - \frac{4(n - 1)}{n} \right] \cdot \tfrac{2}{n} = \left[5 - \frac{4(n - 1)}{n} \right] \cdot \tfrac{2}{n},$$

whose sum is

$$S_n = \left[5n - \left(\tfrac{4}{n} + \tfrac{8}{n} + \ldots + \frac{4(n - 1)}{n} \right) \right] \cdot \tfrac{2}{n}$$

$$= \left[5n - \tfrac{4}{n} (1 + 2 + \ldots + (n - 1)) \right] \cdot \tfrac{2}{n}$$

$$= \left[5n - \tfrac{4}{n} \cdot \frac{(n - 1)n}{2} \right] \cdot \tfrac{2}{n} = 10 - \frac{4(n - 1)}{n} \ .$$

Therefore, $\lim\limits_{n\to\infty} S_n = \lim\limits_{n\to\infty} (10 - 4 + \tfrac{4}{n}) = 6$.

7. $\int_1^4 x\sqrt{x^2 + 1} \ dx = \tfrac{1}{3} (x^2 + 1)^{3/2} \big|_1^4 = \tfrac{1}{3} (17^{3/2} - 2^{3/2}) \approx 22.42146$.

8. (a) $\int_1^4 \frac{(x - 2)^2}{\sqrt{x}} \ dx = \int_1^4 \frac{x^2 - 4x + 4}{\sqrt{x}} \ dx$

$$= \int_1^4 (x^{3/2} - 4x^{1/2} + 4x^{-1/2}) dx$$

$$= \tfrac{2}{5} x^{5/2} - \tfrac{8}{3} x^{3/2} + 8x^{1/2} \big|_1^4$$

$$= (\tfrac{2}{5}(32) - \tfrac{8}{3}(8) + 16) - (\tfrac{2}{5} - \tfrac{8}{3} + 8)$$

$$\approx 1.73333.$$

(b) $\int_{-2}^{0} x^2(4 - x)\,dx = \int_{-2}^{0} (4x^2 - x^3)\,dx = \frac{4}{3}x^3 - \frac{1}{4}x^4\Big|_{-2}^{0}$

$$= -\left[\frac{4}{3}(-2)^3 - \frac{1}{4}(-2)^4\right] = \frac{44}{3} \approx 14.66667.$$

9. $\frac{d}{dx}\int_{1-x^2}^{0} \sqrt[3]{t^2 + 1}\,dt = -\frac{d}{dx}\int_{0}^{1-x^2} \sqrt[3]{t^2 + 1}\,dt$

$$= -\sqrt[3]{\left(1 - x^2\right)^2 + 1} \cdot \frac{d}{dx}(1 - x^2)$$

$$= 2x(x^4 - 2x^2 + 2)^{1/3}.$$

10. Subdivision points are $x_0 = 0$, $x_1 = 1/4$, $x_2 = 1/2$, $x_3 = 3/4$, $x_4 = 1$ and $h = (1 - 0)/4 = 1/4$. Then,
$y_0 = \sqrt{1 + 0} = 1$, $y_1 = \sqrt{1 + (1/64)} \approx 1.00778$,
$y_2 = \sqrt{1 + (1/8)} \approx 1.06066$, $y_3 = \sqrt{1 + (27/64)} \approx 1.19242$,
$y_4 = \sqrt{1 + 1} \approx 1.41421$. Thus,

$$\int_{0}^{1} \sqrt{1 + x^3}\,dx \approx T = \frac{1}{8}(y_0 + 2y_1 + 2y_2 + 2y_3 + y_4)$$

$$\approx \frac{1}{8}(8.93593 \approx 1.11699.$$

11. Subdivision points are $x_0 = 1$, $x_1 = 7/6$, $x_2 = 4/3$, $x_3 = 3/2$, $x_4 = 5/3$, $x_5 = 11/6$, $x_6 = 2$, and $h = 1/6$. Then, $y_0 = 1$,
$y_1 = 6/7 \approx .85714$, $y_2 = 3/4 = .75$, $y_3 = 2/3 \approx .66667$,
$y_4 = 3/5 = .6$, $y_5 = 6/11 \approx .54545$, $y_6 = 1/2 = .5$. Thus,

$$\int_{1}^{2} \frac{dx}{x} \approx \frac{1}{3 \cdot 6}\left[y_0 + 4y_1 + 2y_2 + 4y_3 + 2y_4 + 4y_5 + y_6\right]$$

$$\approx \frac{1}{18}(12.47706) \approx 0.69317.$$

To estimate the error, $f(x) = \frac{1}{x}$, $f'(x) = -\frac{1}{x^2}$, $f^{(3)}(x) = \frac{2}{x^3}$, and $f^{(4)}(x) = -\frac{6}{x^4}$. Since $\left|-\frac{6}{x^4}\right| < 6$ on $1 \le x \le 2$, the error satisfies

$$E = \left|\frac{2 - 1}{180} f^{(4)}(c) \cdot h^4\right| < \frac{1}{180} \cdot 6 \cdot \frac{1}{1296} \approx .00003.$$

Therefore, the approximation is accurate to 4 decimal places.

12. $\int_{0}^{\pi} f(x)\,dx = \int_{0}^{\pi/2} \sin x\,dx + \int_{\pi/2}^{\pi} \pi x\,dx$

$$= -\cos x\,\Big]_{0}^{\pi/2} + \frac{1}{2}\pi x^2\,\Big]_{\pi/2}^{\pi} = 1 + \frac{3\pi^3}{8} \approx 12.62735.$$

CHAPTER 5 APPLICATIONS OF DEFINITE INTEGRALS

INTRODUCTION.

Let us review the central theme of the previous chapter.

1. Let $A(x) = \int_a^x f(t)\ dt$ for the continuous function $y = f(x)$ over the interval $a \leq x \leq b$. Then $A'(x) = $ _____.

2. More generally, this motivated the idea of the definite integral as satisfying the equation
$$\int_a^b f(x)\ dx = \underline{\hspace{6cm}},$$
where $F(x) = \int f(x)\ dx$ is any integral of f.

3. The connection between the definite integral $\int_a^b f(x)\ dx$ and the sums of the form $S_a^b = \sum_a^b f(x)\ \Delta x$ was revealed to be that the definite integral is the _____ of S_a^b as Δx _____, provided that the function f is assumed to be _____ on $a \leq x \leq b$.

4. Thus, a sum of the form S_a^b for suitably small Δx could be used as an approximation to the value of the definite integral $\int_a^b f(x)\ dx$. Two numerical methods providing such an approximation studied in the previous chapter were the _____ rule and _____ rule.

5-1 THE NET CHANGE IN POSITION AND DISTANCE TRAVELED BY A MOVING BODY.

OBJECTIVE A : Given a continuous function $v = f(t)$ representing the velocity v of a moving body as a function of time t, find the time intervals in which the velocity is positive and in which it is negative, and calculate the total distance traveled by the body during the specified time interval $t = a$ to $t = b$.

5. Consider the velocity given by $v = \sin(3t - 1)$ for $0 \leq t \leq 2$. A graph of the velocity is shown in the figure on the next page. From our analysis of the sine curve in Article

---

1. $f(x)$

2. $F(b) - F(a)$

3. limit, approaches zero, continuous

4. trapezoidal, Simpson's

2-6, we observe that the velocity is of periodicity _____,
amplitude _____, and horizontal shift _____.
Thus, the velocity v is
negative for $0 \le t \le 1/3$,
it is positive for _____,
and negative again for
_____.   Therefore,
the total distance traveled by
the moving body is

$s = \int_0^{1/3} - \sin (3t - 1) \, dt$

$+$ _____

$= \frac{1}{3} + \cos (3t - 1)]_0^{1/3}$
$+$ _____

$= \frac{1}{3} (1 - \cos 1) +$ _____

$=$ _____ $\approx 1.24779.$

v = sin (3t – 1)

$\frac{1}{3} + \frac{2\pi}{3} \approx 2.42773$

OBJECTIVE B :   Given a continuous function $a = f(t)$ representing the
acceleration   a   of a moving body as a function of
time   t, and given v_0 as its velocity at time $t = 0$,
find the total distance traveled by the body during
the specified time interval $t = 0$ to $t = b$.

6.   Suppose the brakes are applied to a car traveling 50 mph and
the brakes give the car a constant negative acceleration of
20 ft/sec^2. How far will the car travel before stopping?
Solution. To make the units consistent in the problem, we
convert miles per hour to feet per second. Since 15 mph is
equivalent to 22 ft/sec, it follows that 50 mph is
equivalent to _____ ft/sec. The problem gives $a =$ _____,
so $v = \int a \, dt =$ _____ $+ C$. Since $v_0 = \frac{220}{3}$ when $t = 0$,
it follows that $C =$ _____. Therefore, $v =$ _____.
Next, the total distance traveled after T seconds is given by

$s = \int_0^T \left(-20t + \frac{220}{3}\right) dt =$ _____. When the car

stops, $v =$ _____, or $-20T + \frac{220}{3} = 0$, and this occurs at
time $T =$ _____ seconds. The total distance traveled during
this time is therefore

$s]_{T=\frac{11}{3}} = -10($_____$)^2 + \frac{220}{3}($_____$) =$ _____ ft.

---

5. $\frac{2\pi}{3}$, 1, $\frac{1}{3}$, $\frac{1}{3} < t < \frac{1}{3} + \frac{\pi}{3}$, $\frac{1}{3} + \frac{\pi}{3} < t \le 2$, $\int_{\frac{1}{3}}^{\frac{1}{3}+\frac{\pi}{3}} \sin (3t - 1) \, dt + \int_{\frac{1}{3}+\frac{\pi}{3}}^{2} - \sin (3t - 1) \, dt,$

$-\frac{1}{3} \cos (3t - 1)]_{\frac{1}{3}}^{\frac{1}{3}+\frac{\pi}{3}} + \frac{1}{3} \cos (3t - 1)]_{\frac{1}{3}+\frac{\pi}{3}}^{2}$, $\frac{2}{3} + \frac{1}{3}(1 + \cos 5)$, $\frac{4}{3} + \frac{1}{3} \cos 5 - \frac{1}{3} \cos 1$

6. $\frac{50 \cdot 22}{15} = \frac{220}{3}$, -20, $-20t$, $\frac{220}{3}$, $-20t + \frac{220}{3}$, $-10T^2 + \frac{220}{3} T$, 0, $\frac{11}{3}$, $\frac{11}{3}$, $\frac{11}{3}$, $\frac{1210}{9} \approx 134.44$

5-2   AREA BETWEEN CURVES.

OBJECTIVE A :   Find the area bounded by two given continuous curves
$y_1 = f_1(x)$ and $y_2 = f_2(x)$ over an interval
$a \leq x \leq b$.   It may be required to calculate a and b.

7.   Find the area between the curves $y = 1$ and $y = 1 - x^{-2}$ for
$1 \leq x \leq 4$.

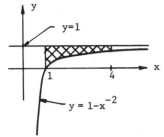

Solution.   The desired region is
shown in the figure at the right.
The area is given by

$$A = \int_1^4 (\underline{\hspace{2cm}}) \, dx = \int_1^4 \underline{\hspace{1.5cm}} \, dx$$

$$= \underline{\hspace{2cm}}]_1^4 = -\frac{1}{4} - (\underline{\hspace{1.5cm}})$$

$$= \underline{\hspace{2cm}}.$$

8.   Find the area between the curves $y = \sin x$ and $y = \frac{2x}{\pi}$.

Solution.   The region is shown in
the figure at the right.   We need
to determine the x-coordinates of
the points   A   and   B   of intersection of the two curves.   Now,
these occur where $\sin x = \frac{2x}{\pi}$;

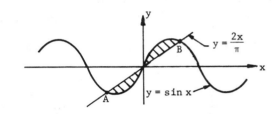

so $x = -\frac{\pi}{2}$, \underline{\hspace{1cm}}, or \underline{\hspace{1cm}}.
As   x   varies from $-\frac{\pi}{2}$ to   0
the curve $y = \sin x$ is \underline{\hspace{1.5cm}} the curve $y = 2x/\pi$;   however,
as   x   varies from   0   to \underline{\hspace{1.5cm}} the situation is reversed.
Thus, the area between the curves is given by

$$A = \int_{-\pi/2}^0 \left(\frac{2x}{\pi} - \sin x\right) dx + \int_{\underline{}}^{\underline{}} (\underline{\hspace{2cm}}) \, dx$$

$$= (\underline{\hspace{2cm}})]_{-\pi/2}^0 + \left(- \cos x - \frac{x^2}{\pi}\right)]_0^{\pi/2}$$

$$= (\underline{\hspace{2cm}}) + \left(- \cos \frac{\pi}{2} - \frac{\pi^2}{4\pi} + 1\right) = \underline{\hspace{1.5cm}}.$$

OBJECTIVE B :   Find the area of the region bounded on the right by the
continuous curve $x_1 = g_1(y)$ and on the left by the
continuous curve $x_2 = g_2(y)$ as   y   varies from $y = c$
to $y = d$.   The calculation of   c   and   d   may be
required.

9.   Find the area bounded by the curves $y^2 = x$ and $y = 6 - x$.
Solution.   The first curve is $x = y^2$
and the second is $x = 6 - y$.   The

---

7. $1 - \left(1 - x^{-2}\right)$, x^{-2}, $-x^{-1}$, -1, $\frac{3}{4}$

8. 0, $\frac{\pi}{2}$, below, $\frac{\pi}{2}$, $\int_0^{\pi/2} \left(\sin x - \frac{2x}{\pi}\right) dx$, $\frac{x^2}{\pi} + \cos x$, $1 - \frac{\pi^2}{4\pi} - \cos\left(-\frac{\pi}{2}\right)$, $2 - \frac{\pi}{2}$

region is depicted in the figure at
the right. First we will find the
y-coordinates of the points A   and
B of intersection of the two
curves. The intersection occurs
when $x = y^2$ is the same as
$x = 6 - y$ or _____.
Equivalently, $y^2 + y - 6 = 0,$ or
$y =$ _____ or $y =$ _____.
Therefore, the area of the region
is given by

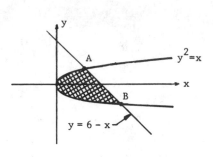

$$A = \int_{-3}^{2} (\text{_____}) \, dy = \text{_____}]_{-3}^{2}$$

$$= (12 - 2 - \tfrac{8}{3}) - (\text{_____}) = \text{_____} \approx \text{_____}.$$

5-3   CALCULATING VOLUMES BY SLICING, VOLUMES OF REVOLUTION.

OBJECTIVE A :   Calculate the volume of a solid of revolution generated
by the graph $y = f(x)$ of a continuous function over
$a \le x \le b$ rotated about the x-axis.

10.   Calculate the volume of the solid generated when $y = x^2 - 6x$
is rotated about the x-axis for $0 \le x \le 6.$

Solution. A graph is shown in the
figure at the right. Thus, the
volume generated is given by

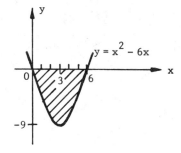

$$V = \text{_____}$$

$$= \int_{0}^{6} \pi\left(x^4 - 12x^3 + 36x^2\right) \, dx$$

$$= \text{_____}$$

$$= \pi(\text{_____}) - \pi \cdot 0 = \text{_____} \approx \text{_____ cubic units.}$$

11.   For the volume generated by the graph of $y = |x|$ rotated
about the x-axis from $x = -2$ to $x = 1,$

$$V = \int_{-2}^{1} \pi(\text{_____})^2 \, dx = \int_{-2}^{1} \text{_____}$$

$$= \pi[\text{_____}]_{-2}^{1} = \pi(\text{_____}) = \text{_____} \approx \text{_____ cubic units.}$$

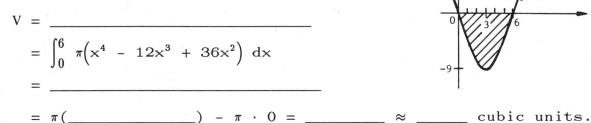

9. $y^2 = 6 - y$, 2, -3, $6 - y - y^2$, $6y - \tfrac{1}{2}y^2 - \tfrac{1}{3}y^3$, $-18 - \tfrac{9}{2} + 9$, $20\tfrac{5}{6}$, 20.83333

10. $\int_{0}^{6} \pi(x^2 - 6x)^2 \, dx$, $\pi[\tfrac{1}{5}x^5 - 3x^4 + 12x^3]_{0}^{6}$, $\tfrac{1}{5}(6)^5 - 3(6)^4 + 12(6)^3$, $\tfrac{1296\pi}{5}$, 814.3

11. $|x|$, $\pi x^2 \, dx$, $\tfrac{1}{3}x^3$, $\tfrac{1}{3} - \left(\tfrac{-8}{3}\right)$, 3π, 9.42

OBJECTIVE B : Use the method of slicing to calculate the volume of a solid whose base is given and whose cross sectional areas are specified.

12. Consider the solid whose base is a triangle cut from the first quadrant by the line $x + 5y = 5$ and whose cross sections perpendicular to the x-axis are semicircles. A typical cross section of the solid is illustrated at the right. From the equation $x + 5y = 5$ we find that $y = $ _____. Since y is the diameter of the semicircle, the cross sectional area given by the area of the semicircle is

$A(x) = \frac{\pi}{2} \left(\underline{\hspace{1.5cm}} \right)^2 = \underline{\hspace{2cm}}.$

Therefore, the volume of the solid is given by

$$V = \int_0^5 A(x)\,dx = \int_0^5 \frac{1}{8}\,\pi y^2\,dx = \int_0^5 \frac{\pi}{8}\left(\underline{\hspace{2cm}}\right)dx$$

$$= \frac{\pi}{8}\left[\frac{x^3}{75} - \frac{x^2}{5} + x\right]_0^5 = \frac{\pi}{8}\left(\underline{\hspace{2.5cm}}\right)$$

$$= \underline{\hspace{1.5cm}} \approx \underline{\hspace{2cm}} \text{ cubic units.}$$

5-4   VOLUMES MODELED WITH WASHERS AND CYLINDRICAL SHELLS.

OBJECTIVE A : Use the method of washers to find the volume generated when a given planar region is rotated about a specified axis or line.

13. Let V denote the volume generated when the planar region bounded by the graph of $y^2 - x^2 - 0$, $y = 0$, and $x = 5$ is rotated about the y-axis. The planar region is illustrated in the figure at the right. Now, imagine the solid to be cut into thin slices by planes perpendicular to the y-axis. Each slice is like a thin washer of thickness Δy, inner radius $r_1 = x = $ _____, and outer radius $r_2 = $ _____. The area of the face of such a washer is $\pi r_2^2 - \pi r_1^2 = \pi\left(\underline{\hspace{1.5cm}}\right)$. Therefore, the volume of the solid is given by

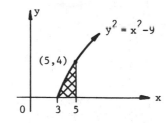

$$V = \int_0^4 \underline{\hspace{2cm}}\,dy = \pi\left[\underline{\hspace{1.5cm}}\right]_0^4 = \underline{\hspace{1.5cm}} \text{ cubic units.}$$

---

12. $-\frac{1}{5}x + 1$, $\frac{y}{2}$, $\frac{1}{8}\pi y^2$, $\frac{x^2}{25} - \frac{2x}{5} + 1$, $\frac{5}{3} - 5 + 5$, $\frac{5\pi}{24}$, 0.655

13. $\sqrt{y^2 + 9}$, 5, $16 - y^2$, $\pi\left(16 - y^2\right)$, $16y - \frac{1}{3}y^3$, $\frac{128\pi}{3}$

14.   Consider the volume generated by rotating the planar region
bounded by $y = x^2$ and $y = |x| + 2$
about the y-axis. A sketch of the
region is shown at the right. The
two curves intersect at the points
where $y = x^2$ is the same as
$y = |x| + 2$. For $x > 0$, this
occurs when $x^2 = x + 2$ or
$x =$ _____. From symmetry it
follows that the points of intersection are _____ and _____.
Imagine the solid to be cut into
thin slices by planes perpendicular
to the y-axis. As y varies from
$y = 0$ to $y = 2$, each slice is a thin washer of thickness
Δy, inner radius 0, and outer radius $x =$ _____. As y
varies from $y = 2$ to $y = 4$, each slice is a thin washer of
thickness Δy, inner radius $x =$ _____ and outer radius
$x = \sqrt{y}$. Therefore, the volume is given by

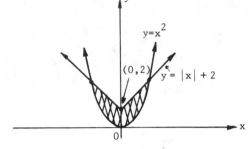

$$V = \int_0^2 \pi(y - \underline{\hspace{1cm}})\, dy + \int_2^4 \pi[\underline{\hspace{2cm}}]\, dy$$

$$= \frac{\pi y^2}{2}\Big]_0^2 + \pi\left(\underline{\hspace{2cm}}\right)\Big]_2^4 = 2\pi + \pi\left[\left(8 - \frac{8}{3}\right) - \left(\underline{\hspace{2cm}}\right)\right]$$

$$= \underline{\hspace{3cm}} \approx 16.76 \text{ cubic units.}$$

OBJECTIVE B :   Use the method of cylindrical shells to find the volume
generated when a given planar region is rotated about a
specified axis.

15.   Consider the volume generated when
the region between the curves $y = x^2$
and $y^2 = x$ is rotated about the
x-axis. The planar area is shown in
the figure at the right. Using the
method of shells, we begin with a
horizontal strip of width Δy above
the x-axis. A typical length of this
strip is $x = \sqrt{y}$ minus $x = y^2$, or
in terms of y, _____. The
volume generated by revolving this strip about the x-axis is a
hollow cylindrical shell of inner circumference _____, inner
length _____, and wall thickness ___. Hence, the total
volume is given by the definite integral

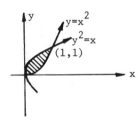

$$V = \int_0^1 \underline{\hspace{3cm}} dy = 2\pi \int_0^1 (\underline{\hspace{2cm}})\, dy$$

$$= 2\pi\ [\underline{\hspace{2cm}}]_0^1 = 2\pi\left(\frac{2}{5} - \frac{1}{4}\right) = \underline{\hspace{1.5cm}} \approx 0.94248 \text{ cubic units.}$$

---

14. 2, (-2,4), (2,4), $\sqrt{y}$, y - 2, 0, y - (y -2)^2, $\frac{y^2}{2}$ - $\frac{(y-2)^3}{3}$, 2 - 0, $\frac{16\pi}{3}$

15. $\sqrt{y}$ - y^2, $2\pi y$, $\sqrt{y}$ - y^2, Δy, $2\pi y\left(\sqrt{y} - y^2\right)$, $y^{3/2}$ - y^3, $\frac{2}{5}$ y$^{5/2}$ - $\frac{1}{4}$ y^4, $\frac{3\pi}{10}$

16. Compute the volume generated by
     rotating about the y-axis the
     region bounded by $y = 2x$ and
     $y = x^2 - 2x$.

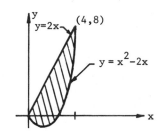

     Solution. A graph of the plane
     region to be rotated is shown in
     the figure at the right. We begin
     with a horizontal strip of width
     Δx. A typical height of this
     strip is $y = 2x$ minus
     $y = $ _____.   The volume
     generated by revolving this strip about the y-axis is a hollow
     cylindrical shell of inner circumference _____, inner length
     _____, and thickness _____.   Hence, the total volume is
     given by the definite integral

     $$V = \int_0^4 \underline{\hspace{2cm}} \, dx = 2\pi \int_0^4 \underline{\hspace{2cm}} \, dx$$

     $$= 2\pi\Big[\underline{\hspace{2cm}}\Big]_0^4 = 128\pi\Big(\frac{4}{3} - \underline{\hspace{1.5cm}}\Big)$$

     $$= \underline{\hspace{2cm}} \approx 134.04 \quad \text{cubic units.}$$

5-5   LENGTHS OF PLANE CURVES.

OBJECTIVE A :   Find the length of a planar curve between two specified
                points, when the curve is defined by an equation that
                gives y as a continuously differentiable function of
                x, or gives x as a continuously differentiable
                function of y.

17. Consider the curve defined by the function $y = \frac{2}{3}(x - 1)^{3/2}$ for
     $1 \leq x \leq 5$. We seek the length of the graph from $(1,0)$ to
     $(5, \frac{16}{3})$. $\frac{dy}{dx} = $ _____, so that $1 + \Big(\frac{dy}{dx}\Big)^2 = $ _____.   The
     arc length is given by

     $$s = \int_1^5 \sqrt{1 + \Big(\frac{dy}{dx}\Big)^2} \, dx = \int_1^5 \underline{\hspace{2cm}} \, dx$$

     $$= \underline{\hspace{2cm}}\Big]_1^5 = \underline{\hspace{2cm}} \approx 6.78689 \quad \text{units.}$$

18. Let the curve be defined by the equation $y^2 = x^3$. To find the
     arc length from the point $(0,0)$ to the point $(4,8)$, we
     first calculate dy/dx. Differentiating the equation for the
     curve implicitly gives

     $2y \frac{dy}{dx} = $ _____ so that $1 + \Big(\frac{dy}{dx}\Big)^2 = 1 + \dfrac{\underline{\hspace{1cm}}}{4y^2} = $ _____.

---

16. $x^2 - 2x$, $2\pi x$, $2x - (x^2 - 2x)$, Δx, $2\pi x(4x - x^2)$, $4x^2 - x^3$, $\frac{4}{3}x^3 - \frac{1}{4}x^4$, 1, $128\pi/3$

17. $\sqrt{x - 1}$, x, $\sqrt{x}$, $\frac{2}{3}x^{3/2}$, $\frac{2}{3}\Big(5\sqrt{5} - 1\Big)$

Therefore, the arc length is given by

$$s = \int \underline{\hspace{2cm}} dx = \frac{4}{9} \cdot \frac{2}{3} \left[\underline{\hspace{3cm}}\right]_0^4$$

$$= \frac{8}{27} (\underline{\hspace{2.5cm}}) = \approx 9.073 \quad \text{units.}$$

OBJECTIVE B :   Find the length of a curve specified parametrically by continuously differentiable equations $x = f(t)$ and $y = g(t)$ over a given interval $a \leq t \leq b$.

19.   To compute the length of the curve given by $x = t^2 \cos t$ and $y = t^2 \sin t$ for $0 \leq t \leq 1$, we first calculate dx/dt and dy/dt.

$$\frac{dx}{dt} = \underline{\hspace{3cm}} ; \quad \frac{dy}{dt} = \underline{\hspace{3cm}} \quad \text{so that}$$

$$\left(\frac{dx}{dt}\right)^2 + \left(\frac{dy}{dt}\right)^2 = t^2[(2 \cos t - t \sin t)^2 + (\underline{\hspace{3cm}})^2]$$

$$= t^2[4 \cos^2 t - 4t \cos t \sin t + t^2 \sin^2 t$$

$$+ (\underline{\hspace{4cm}})]$$

$$= t^2[4 + \underline{\hspace{1.5cm}}].$$

Hence the arc length is given by,

$$s = \int_0^1 \sqrt{\left(\frac{dx}{dt}\right)^2 + \left(\frac{dy}{dt}\right)^2} \, dt = \int_0^1 \underline{\hspace{3cm}} \, dt$$

$$= \underline{\hspace{2.5cm}}]_0^1 = \frac{1}{3} (\underline{\hspace{2cm}}) \approx 1.0601 \quad \text{units.}$$

5-6   THE AREA OF A SURFACE OF REVOLUTION.

OBJECTIVE A :   Find the area of a surface generated by rotating the portion of a given curve $y = f(x)$ between $x = a$ and $x = b$ about a specified axis or line parallel to an axis. The function f is assumed to have a derivative that is continuous on the interval.

20.   Find the surface area when the arc $y = 2\sqrt{x}$ from $x = 0$ to $x = 8$ is rotated about the x-axis.
      Solution.   We write $dS = 2\pi y \, ds = 2\pi \cdot 2\sqrt{x} \cdot \underline{\hspace{2.5cm}} \, dx.$
      $\frac{dy}{dx} = \underline{\hspace{1.5cm}}$ so that $\sqrt{1 + \left(\frac{dy}{dx}\right)^2} = \underline{\hspace{3cm}}.$
      Thus,

---

18. $3x^2$, $9x^4$, $1 + \frac{9x}{4}$, $\int_0^4 \sqrt{1 + \frac{9x}{4}} \, dx$, $\left(1 + \frac{9x}{4}\right)^{3/2}$, $10\sqrt{10} - 1$

19. $2t \cos t - t^2 \sin t$, $2t \sin t + t^2 \cos t$, $2 \sin t + t \cos t$, $4 \sin^2 t + 4t \cos t \sin t + t^2 \cos^2 t$,

      t^2, $t\sqrt{4 + t^2}$, $\frac{1}{3}\left(4 + t^2\right)^{3/2}$, $5\sqrt{5} - 8$

$$S = \int_{\underline{}}^{\underline{}} \underline{\hspace{4cm}} \, dx = \int_{\underline{}}^{\underline{}} \underline{\hspace{4cm}} \, dx$$

$$= 4\pi\left(\underline{\hspace{4cm}}\right)]_0^8 = \frac{8\pi}{3}\left(\underline{\hspace{2cm}} - 1\right)$$

$$= \underline{\hspace{2cm}} \approx 217.81709 \quad \text{sq. units.}$$

21. Find the surface area generated when the arc $y = x^{1/3}$ from $x = 0$ to $x = 1$ is rotated about the y-axis.

    Solution. We write $dS = 2\pi\underline{\hspace{1cm}} ds$, since the rotation occurs about the y-axis. Since $x = y^3$, $dx/dy = \underline{\hspace{2cm}}$ and

    $$ds = \sqrt{1 + \left(\frac{dy}{dx}\right)^2} \, dx = \sqrt{1 + \left(\frac{dx}{dy}\right)^2} \, dy = \underline{\hspace{3cm}} \, dy.$$

    At $x = 0$, $y = 0$ and at $x = 8$, $y = 2$ so the surface area is

    $$S = \int 2\pi x \, ds = 2\pi \int_{\underline{}}^{\underline{}} \underline{\hspace{4cm}} \, dy$$

    $$= 2\pi\left(\frac{1}{36} \cdot \frac{2}{3}\right)\left(\underline{\hspace{3cm}}\right)]_{\underline{}}^{\underline{}} = \frac{\pi}{27}\left(\underline{\hspace{3cm}}\right)$$

    $$\approx 203.0436 \quad \text{sq. units.}$$

OBJECTIVE B : Find the area of a surface generated by rotating the arc of a curve specified parametrically by equations $x = f(t)$ and $y = g(t)$ over $a \leq t \leq b$ about an indicated axis. (Again, f and g are assumed to have continuous derivatives.)

22. Find the surface area generated when the arc $x = 2t$ and $y = \sqrt{2}\, t^2$ from $t = 0$ to $t = 2$ is rotated about the y-axis.

    Solution. We write $dS = 2\pi \underline{\hspace{1cm}} ds$, since the rotation occurs about the y-axis. Next, we calculate the derivatives $dx/dt = \underline{\hspace{2cm}}$ and $dt/dt = \underline{\hspace{2cm}}$ so that the arc length differential is given by

    $$ds = \sqrt{\underline{\hspace{3cm}}} \, dt = 2\sqrt{\underline{\hspace{3cm}}} \, dt.$$

    Therefore, the surface area is

    $$S = 2\pi \int_0^2 \underline{\hspace{3cm}} \, dt = \frac{4\pi}{3}\left(\underline{\hspace{3cm}}\right)]_0^2$$

    $$= \frac{4\pi}{3}\left(\underline{\hspace{2cm}}\right) = \underline{\hspace{3cm}} \approx 108.90854 \quad \text{sq. units.}$$

5-7   THE AVERAGE VALUE OF A FUNCTION.

OBJECTIVE : Calculate the average value of a given continuous function $y = f(x)$ over a specified interval $a \leq x \leq b$.

---

20. $\sqrt{1 + \left(\frac{dy}{dx}\right)^2}$, $x^{-1/2}$, $\sqrt{1 + x^{-1}}$, $\int_0^8 4\pi\sqrt{x}\sqrt{1 + x^{-1}} \, dx$, $\int_0^8 4\pi\sqrt{x + 1} \, dx$, $\frac{2}{3}(x + 1)^{3/2}$, $9^{3/2}$, $\frac{208\pi}{3}$

21. x, $3y^2$, $\sqrt{1 + 9y^4}$, $\int_0^2 y^3\sqrt{1 + 9y^4} \, dy$, $\left(1 + 9y^4\right)^{3/2}]_0^2$, $(145)^{3/2} - 1$

22. x, 2, $2\sqrt{2}\, t$, $\left(\frac{dx}{dt}\right)^2 + \left(\frac{dy}{dt}\right)^2$, $2t^2 + 1$, $4t\sqrt{2t^2 + 1}$, $\left(2t^2 + 1\right)^{3/2}$, $27 - 1$, $\frac{104\pi}{3}$

23.   To find the average value of $y = \sin x - \cos x$ over
$0 \le x \le \pi/4$ we have

$$\left(y_{av}\right)_x = \frac{1}{\frac{\pi}{4} - 0} \int_{\underline{\quad}}^{\overline{\quad}} (\underline{\hspace{3cm}}) \, dx$$

$$= \frac{4}{\pi} (\underline{\hspace{2.5cm}})]_{\underline{\quad}}^{\overline{\quad}}$$

$$= \frac{4}{\pi}\left[\left(-\frac{\sqrt{2}}{2} - \frac{\sqrt{2}}{2}\right) - (\underline{\hspace{1.5cm}})\right] = \frac{4}{\pi} (\underline{\hspace{1.5cm}}) \approx -.52739.$$

24.   Suppose that $y = f(x)$ has a continuous derivative over the
interval $a \le x \le b$. Since the slope   m   of the tangent line
at any point is $f'(x)$,   the average value of the slope is given
by

$$\left(m_{av}\right)_x = \underline{\hspace{4cm}} = \frac{\overline{\hspace{3cm}}}{b - a} \; .$$

Interpreting this last equation geometrically, the average
slope of the tangent line over the interval is precisely the
slope of the chord joining the points _____ and
_____ on the graph of $y = f(x)$.   Do you see how this
motivates the name of the "Mean Value Theorem" of Article 3-7?

5-8   MOMENTS AND CENTERS OF MASS.

OBJECTIVE A :   Find the center of mass of a thin homogeneous plate
covering a given region of the xy-plane.

25.   Find the center of mass $(\overline{x}, \overline{y})$ of the thin homogeneous
triangular plate formed in the first quadrant by the coordinate
axes and the line $x + 2y = 2$.

Solution.   The plate is depicted in
the figure at the right. To find
$\overline{x}$,   divide the triangle into vertical
strips of width   dx   parallel to the
y-axis. A representative strip is
shown in the figure and its mass is
approximately $dm = \delta_2 dA = \delta_2(y \underline{\quad}) = \delta_2(\underline{\quad}) \, dx$.   Thus,

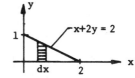

$$\overline{x} = \frac{\int x \, dm}{\int dm} = \frac{\int_0^2 x \cdot (\underline{\quad}) \, dx}{\int_0^2 \delta_2\left(1 - \frac{x}{2}\right) \, dx} = \frac{\int_0^2 (\underline{\quad}) \, dx}{\int_0^2 \left(1 - \frac{x}{2}\right) \, dx}$$

$$= \frac{\underline{\hspace{2cm}}]_0^2}{x - \frac{1}{4} x^2]_0^2} = \frac{\underline{\hspace{1.5cm}}}{2 - 1} = \underline{\hspace{1.5cm}} \; .$$

---

23. $\int_0^{\pi/4} (\sin x - \cos x) \, dx, \quad (-\cos x - \sin x)]_0^{\pi/4}, \quad (-1 - 0), \quad (1 - \sqrt{2})$

24. $\frac{1}{b - a}\int_a^b f'(x) \, dx, \quad f(b) - f(a), \quad \left(a, f(a)\right), \quad \left(b, f(b)\right)$

In a similar manner, to find $\bar{y}$, divide
the triangle into horizontal strips of
width dy. The mass of a representative
strip is approximately

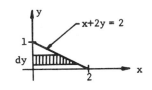

$$dm = \delta_2 dA = \delta_2 (\underline{\quad} \, dy) = \delta_2 (\underline{\qquad}) \; dy.$$
Thus,

$$\bar{y} = \frac{\int y \; dm}{\int dm} = \frac{\int_0^1 y \cdot (\underline{\qquad}) \; dy}{\int_0^1 \delta_2 (2 - 2y) \, dy} = \frac{\int_0^1 (\underline{\qquad}) \; dy}{\int_0^1 (1 - y) \, dy}$$

$$= \frac{\underline{\qquad} \Big]_0^1}{2y - y^2 \; \Big]_0^1} = \frac{\underline{\qquad}}{2 - 1} = \underline{\qquad}.$$

$\boxed{\text{OBJECTIVE B}}$:   Find the center of gravity of the planar region bounded
by given curves and lines.

26.   Find the center of gravity of the region bounded by the
parabola $y = x^2$ and the line $y = x$.
Solution. The region is illustrated
at the right. To calculate $\bar{x}$, divide
the region into vertical strips of width
dx. The area of a typical strip is
given by $dA = (\underline{\qquad}) \; dx$ so that

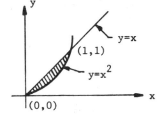

$$\bar{x} = \frac{\int x \; dA}{\int dA} = \frac{\int_0^1 \underline{\qquad} \; dx}{\int_0^1 (x - x^2) \; dx}$$

$$= \frac{\underline{\qquad} \Big]_0^1}{\frac{1}{2} x^2 - \frac{1}{3} x^3 \Big]_0^1} = \frac{\underline{\qquad}}{\frac{1}{2} - \frac{1}{3}} = \frac{\underline{\qquad}}{\frac{1}{6}} = \underline{\qquad}.$$

To find $\bar{y}$, divide the region into horizontal strips of width
dy. The area of a typical strip is $dA = (\underline{\qquad}) \; dy$ so that

$$\bar{y} = \frac{\int y \; dA}{\int dA} = \frac{\underline{\qquad}}{\frac{1}{6}} = \frac{\underline{\qquad} \Big]_0^1}{\frac{1}{6}}$$

$$= 6 (\underline{\qquad}) = 6 (\underline{\qquad}) = \underline{\qquad}.$$

---

25. dx, $1 - \frac{x}{2}$, $\delta_2 \left(1 - \frac{x}{2}\right)$, $x - \frac{1}{2} x^2$, $\frac{1}{2} x^2 - \frac{1}{6} x^3$, $\frac{2}{3}$, $\frac{2}{3}$, x, $2 - 2y$, $\delta_2 (2 - 2y)$, $2(y - y^2)$, $y^2 - \frac{2}{3} y^3$, $1 - \frac{2}{3}$, $\frac{1}{3}$

26. $x - x^2$, $x(x - x^2)$, $\frac{1}{3} x^3 - \frac{1}{4} x^4$, $\frac{1}{3} - \frac{1}{4}$, $\frac{1}{12}$, $\frac{1}{2}$, $\sqrt{y} - y$, $\int_0^1 y \left(\sqrt{y} - y\right) dy$, $\frac{2}{5} y^{5/2} - \frac{1}{3} y^3$, $\frac{2}{5} - \frac{1}{3}$,

$\frac{1}{15}$, $\frac{2}{5}$

5-9 WORK.

OBJECTIVE A :   For a mechanical spring of given natural length L, if
a specified force is required to stretch or compress
the spring by a certain given amount, calculate (a) the
"spring constant" c and (b) the amount of work done
in stretching or compressing the spring from a
specified length L = a to a specified length L = b.

27.   Suppose an unstretched spring is 3 feet long. When the
spring is used to suspend a 4-pound weight, it stretches to a
length of 5 feet. Therefore, the equation F = cs becomes
_____, or c = _____ is the value of the spring
constant. To calculate the work required to stretch the spring
from a length of 4 feet to a length of 6 feet, we find

$$W = \int F \, ds = \int_1^{\overline{}} (\underline{}) \, dx = \underline{} \text{ foot-pounds.}$$

OBJECTIVE B :   Find the work done in pumping all the water out of a
specified container to a given height above the
container.

28.   A cylindrical tank of radius 5 feet and height 10 feet
stands on a platform so that its bottom is 50 feet above the
surface of a lake. Water is pumped directly up from the
surface of the lake into the water tank. Find the depth d of
water in the tank when half the necessary work has been done to
fill the tank.

Solution. Let the distance from the
bottom of the tank to the surface of
the water in the tank be x feet.
The situation is depicted in the
figure at the right. Consider a
slice of water in the tank of thickness dx cut by two planes parallel
to the surface of the lake and
platform. The volume of this slice is
dV = A dx = _____. Therefore, its
weight is w dV = _____, where w
is the weight of a cubic foot of water.
The total distance this slice is lifted
is 50 + x feet, where x is its
typical height from the bottom of the
tank. Therefore the total work required
to fill the tank is

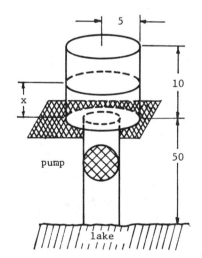

$$W_1 = \int \text{distance} \cdot w \, dV = \int_0^{10} (\underline{}) \, dx$$

$$= 25\pi w \, (\underline{})]_0^{10} = 25\pi w (\underline{}) .$$

---

27. $4 = c(5 - 3)$, 2, $\int_1^3 2x \, dx$, 8

On the other hand, the work required to fill the tank to the depth d is given by

$$W_2 = \int_0^d (\underline{\hspace{3cm}}) \, dx = 25\pi w(\underline{\hspace{3cm}}).$$

We want to find d when $W_2 = \frac{1}{2} W_1$. This translates into

$25\pi w(\underline{\hspace{2cm}}) = \frac{1}{2} (25\pi w)(550)$ or $\underline{\hspace{2cm}} = 550$.

Solving this quadratic equation for d (which must be positive) gives

$$d = \frac{-100 \pm \sqrt{10000 + \underline{\hspace{2cm}}}}{2}, \quad \text{or} \quad d \approx \underline{\hspace{2cm}} \text{ feet.}$$

5-10 HYDROSTATIC FORCE.

OBJECTIVE :   Find the total force exerted on a given planar region placed vertically under the surface of an incompressible fluid of constant density w.

29. Suppose the face of a dam has the shape of an isosceles trapezoid of altitude 20 feet with an upper base of 30 feet and a lower base of 20 feet. A figure illustrating the face of the dam is shown at the right. Let us find the force due to water pressure on the dam when the surface of the water is level with the top of the dam.

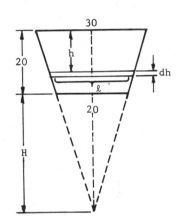

From the figure and similar triangles, we see that $\frac{20}{H} = \frac{30}{\underline{\hspace{1cm}}}$ or

$30H = \underline{\hspace{2cm}}$; thus $H = \underline{\hspace{1cm}}$ feet. Next, consider a horizontal strip of the face of width dh at a depth h below the surface of the water (see figure). Let ℓ denote the horizontal length of this strip. Then by similar triangles, $\frac{\ell}{60 - h} = \frac{30}{\underline{\hspace{1cm}}}$, or $\ell = \underline{\hspace{2cm}}$. Therefore, the area of the strip is $dA = \ell dh$, so the total force exerted on the face of the dam is

$$F = \int wh \, dA = \int wh\ell \, dh = \int_0^{20} \underline{\hspace{2cm}} = \frac{w}{2} \int_0^{20} (\underline{\hspace{2cm}}) \, dh$$

$$= \frac{w}{2} \left[\underline{\hspace{3cm}} \right]_0^{20} = 200w \left(30 - \underline{\hspace{1cm}} \right)$$

$$= \underline{\hspace{3cm}} \approx 291{,}667 \text{ pounds using } w = 62.5.$$

---

28. $25\pi \, dx$, $25\pi w \, dx$, $25\pi w(50 + x)$, $50x + \frac{1}{2} x^2$, 550, $25\pi w(50 + x)$, $50d + \frac{1}{2} d^2$, $50d + \frac{1}{2} d^2$,

$100d + d^2$, 2200, 5.227

29. $20 + H$, $400 + 20H$, 40, 60, $\frac{1}{2}(60 - h)$, $\frac{w}{2} h(60 - h) \, dh$, $60h - h^2$, $30h^2 - \frac{1}{3} h^3$, $\frac{20}{3}$, $\frac{14000w}{3}$

CHAPTER 5   SELF-TEST

1. Find the area of the planar region bounded by the curves
    $y = x^3 + 1$ and $y = x^2 + x$.

2. Find the area of the planar region bounded by the curves $x = y^{1/3}$
    and $y^2 = x$.

3. A train leaving a railroad station has an acceleration of
    $a = 0.5 + 0.02t$ ft/sec^2. How far will the train move in the
    first 20 sec of motion? What is its velocity after 20
    seconds?

4. Find the volume generated by rotating the ellipse $\dfrac{x^2}{a^2} + \dfrac{y^2}{b^2} = 1$
    about the x-axis.

5. The base of a certain solid is the region between the planar
    curves $y = 4 + x^2$ and $y = 12 - x^2$, and the cross sections by
    planes perpendicular to the x-axis are circles with diameters
    extending from one curve to the other. Find the volume of the
    solid.

6. Find the volume generated when the planar region between the
    lines $y = x$, $y = 3x$ and $x + y = 4$ is rotated about the
    x-axis.

7. Find the volume generated when the planar region between the
    curve $y^2 = 9x$ and the lines $x = 4, y = 0$ is rotated about the
    line $x = 5$.

8. Find the lengths of the following curves.

    (a) $y = \dfrac{2}{3} x^{3/2} - \dfrac{1}{2} x^{1/2}$ over $0 \le x \le 1$, and

    (b) $x = t^3 + 3t^2$ and $y = t^3 - 3t^2$ for $0 \le t \le 2$.

9. Find the surface area generated by revolving about the x-axis the
    arc $y = \dfrac{1}{2\sqrt{2}} x\sqrt{1 - x^2}$, $0 \le x \le 1$.

10. Find the area of the surface generated by rotating the cardiod
    $x = 2 \cos \theta - \cos 2\theta$, $y = 2 \sin \theta - \sin 2\theta$, for $0 \le \theta \le \pi$,
    about the x-axis. First sketch the curve.

11. Find the average value of the function $y = x^2 - x + 1$ over the
    interval $0 \le x \le 2$.

12. Find the center of gravity of the planar region bounded by the
    curves $y = x + 2, y = 2x$, and the y-axis.

13. A water pipe is in the shape of a cylinder placed horizontally in
    the ground. Its radius is 4 inches. If the pipe is half full
    of water, find the force on the face of the vertical circular
    plate that closes off the pipe. Assume the density of the water
    is $w = 62.5$ pounds per cubic foot.

14. If a force of 90 pounds stretches a 10-foot spring by 1 foot, find the work done in stretching the spring from 10 feet to 15 feet.

15. A conical tank is 16 feet across the top, and 12 feet deep. It contains water to a depth of 8 feet. Find the work required to pump all the water to a height of 2 feet above the top of the tank.

SOLUTIONS TO CHAPTER 5 SELF-TEST

1. The graph of $y = x^3 + 1$ crosses the x-axis at $(-1,0)$ and so does the graph of $y = x^2 + x$. (See the figure at the right.) Solving for the other point of intersection of the two curves, $x^3 + 1 = x^2 + x$ or $x^3 - x^2 - x + 1 = 0$. Since $x = -1$ is a root, by division $x^3 - x^2 - x + 1 = (x + 1)(x^2 - 2x + 1) = 0$ or $(x + 1)(x - 1)^2 = 0$. Thus, the other point of intersection is $(1,2)$. The area between the curves is then given by

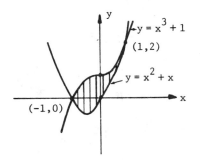

$$A = \int_{-1}^{1} \left[(x^3 + 1) - (x^2 + x) \right] dx = \left(\tfrac{1}{4}x^4 + x - \tfrac{1}{3}x^3 - \tfrac{1}{2}x^2 \right) \Big]_{-1}^{1}$$

$$= \tfrac{4}{3} \text{ square units.}$$

2. The planar region is shown at the right. The points of intersection of the two graphs are $(0,0)$ and $(1,1)$. Thus the area is given by

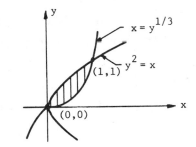

$$A = \int_{0}^{1} \left(y^{1/3} - y^2 \right) dy$$

$$= \tfrac{3}{4}y^{4/3} - \tfrac{1}{3}y^3 \Big]_{0}^{1} = \tfrac{5}{12} \text{ square units.}$$

Alternatively,

$$A = \int_{0}^{1} \left(\sqrt{x} - x^3 \right) dx$$

$$= \tfrac{2}{3}x^{3/2} - \tfrac{1}{4}x^4 \Big]_{0}^{1} = \tfrac{5}{12} \text{ square units.}$$

3. Since $v = \int a \, dt$ we have $v = 0.5t + 0.01t^2 + C_1$. At $t = 0$ the train is at rest, so $v = 0$; hence $C_1 = 0$. Next, $s = \int v \, dt$ or $s = 0.25t^2 + \tfrac{1}{300}t^3 + C_2$. At $t = 0$, $s = 0$ so that $C_2 = 0$. Thus, when $t = 20$ seconds, $s = \tfrac{1}{4}(400) + \tfrac{1}{300}(800) = 126 \; 2/3$ feet, the distance traveled by the train in the first 20 seconds. Its velocity at that time is

$$v = (0.5)(20) + (0.01)(400) = 14 \text{ ft/sec.}$$

4. $V = \int_{-a}^{a} \pi y^2 \, dx = \int_{-a}^{a} \pi \cdot \frac{b^2}{a^2} (a^2 - x^2) \, dx$

$= \frac{\pi b^2}{a^2} [a^2 x - \frac{1}{3} x^3]_{-a}^{a} = \frac{4}{3} \pi b^2 a.$

This generalizes the formula for the volume of a sphere, when
a = b = r   is its radius.

5.   The two curves intersect when
$4 + x^2 = 12 - x^2$ or $x = \pm 2.$
(See the figure at the right
depicting the base of the solid.)
The cross-sectional area of a typical
slice is

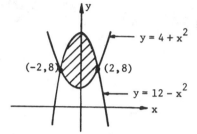

$\pi \left(\frac{12 - x^2 - (4 + x^2)}{2}\right)^2 = \pi \left(4 - x^2\right)^2 = \pi (16 - 8x^2 + x^4).$

Thus the volume is given by

$V = \int A(x) \, dx = \int_{-2}^{2} \pi (16 - 8x^2 + x^4) \, dx$

$= \pi [16x - \frac{8}{3} x^3 + \frac{1}{5} x^5]_{-2}^{2} = \frac{512\pi}{15} \approx 107.2$ cubic units.

6.   The planar region to be rotated about
the x-axis is shown in the figure at
the right. We slice the region into
horizontal slices and use the method
of cylindrical shells. Thus, from
the figure we see that

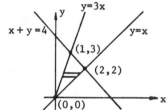

$V = \int_{0}^{2} 2\pi y (y - \frac{y}{3}) dy + \int_{2}^{3} 2\pi y (4 - y - \frac{y}{3}) \, dy$

$= 2\pi [\frac{2}{9} y^3]_{0}^{2} + 2\pi [2y^2 - \frac{4}{9} y^3]_{2}^{3} = \frac{20\pi}{3} \approx 20.9$ cubic units.

7.   The planar region to be rotated is
shown in the figure at the right. We
slice the region into horizontal
slices and use the method of washers.
Thus, from the figure we find the
volume is,

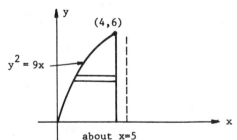

$V = \int_{0}^{6} \pi \left[\left(5 - \frac{y^2}{9}\right)^2 - 1^2\right] dy$

$= \pi \int_{0}^{6} \left(24 - \frac{10}{9} y^2 + \frac{1}{81} y^4\right) dy$

$= \pi [24y - \frac{10}{27} y^3 + \frac{1}{405} y^5]_{0}^{6} = 83.2\pi \approx 261.4$ cubic units.

8.   (a) $y = \frac{2}{3} x^{3/2} - \frac{1}{2} x^{1/2}$ and $\frac{dy}{dx} = x^{1/2} - \frac{1}{4} x^{-1/2}$

$1 + \left(\frac{dy}{dx}\right)^2 = \left(x - \frac{1}{2} + \frac{1}{16} x^{-1}\right) + 1 = \left(x^{1/2} + \frac{1}{4} x^{-1/2}\right)^2 ,$

so the arc length is given by

$s = \int_{0}^{1} \left(x^{1/2} + \frac{1}{4} x^{-1/2}\right) dx = \frac{2}{3} x^{3/2} + \frac{1}{2} x^{1/2}]_{0}^{1} = \frac{7}{6}$ units.

(b) $\dfrac{dx}{dt} = 3t^2 + 6t$ and $\dfrac{dy}{dt} = 3t^2 - 6t$ so that

$$\left(\dfrac{dx}{dt}\right)^2 + \left(\dfrac{dy}{dt}\right)^2 = \left(9t^4 + 36t^3 + 36t^2\right) + \left(9t^4 - 36t^3 + 36t^2\right)$$
$$= 18t^2\left(t^2 + 4\right).$$

Thus, the arc length is given by

$$s = \int_0^2 3\sqrt{2}\; t\sqrt{t^2 + 4}\; dt = 3\sqrt{2}\left(\tfrac{1}{3}\right)\left(t^2 + 4\right)^{3/2}\;\Big]_0^2$$

$$= 8(4 - \sqrt{2}) \approx 20.7 \quad \text{units.}$$

9. Notice that $8y^2 = x^2 - x^4$. Differentiating implicitly,

$$16y\left(\dfrac{dy}{dx}\right) = 2x - 4x^3 \quad \text{or} \quad \dfrac{dy}{dx} = \dfrac{x - 2x^3}{8y}. \quad \text{Thus,}$$

$$1 + \left(\dfrac{dy}{dx}\right)^2 = 1 + \dfrac{\left(x - 2x^3\right)^2}{64y^2} = 1 + \dfrac{\left(x - 2x^3\right)^2}{8\left(x^2 - x^4\right)} = \dfrac{\left(3 - 2x^2\right)^2}{8\left(1 - x^2\right)}\; .$$

Therefore, the surface area is given by

$$S = \int 2\pi y\; ds = \int_0^1 2\pi\; \dfrac{x}{2\sqrt{2}}\; \sqrt{1 - x^2}\cdot \dfrac{3 - 2x^2}{2\sqrt{2}\sqrt{1 - x^2}}\; dx$$

$$= \tfrac{\pi}{4}\int_0^1 \left(3 - 2x^2\right) x\; dx = \tfrac{\pi}{4}[\tfrac{3}{2}x^2 - \tfrac{1}{2}x^4]_0^1 = \tfrac{\pi}{4} \quad \text{square units.}$$

10. The required surface is obtained by rotating the arc from $\theta = 0$ to $\theta = \pi$ about the x-axis. The arc is shown in the figure at the right.
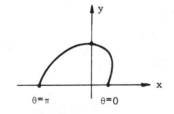
$\dfrac{dx}{d\theta} = -2\sin\theta + 2\sin 2\theta,$

$\dfrac{dy}{d\theta} = 2\cos\theta - 2\cos 2\theta$ so that

$$\left(\dfrac{dx}{d\theta}\right)^2 + \left(\dfrac{dy}{d\theta}\right)^2 = \left(4\sin^2\theta - 8\sin\theta\sin 2\theta + 4\sin^2 2\theta\right) +$$
$$\left(4\cos^2\theta - 8\cos\theta\cos 2\theta + 4\cos^2 2\theta\right)$$

$$= 8(1 - \sin\theta\sin 2\theta - \cos\theta\cos 2\theta)$$

$$= 8[1 - \cos(2\theta - \theta)]$$

$$= 8(1 - \cos\theta)$$

Therefore, the surface area is given by

$$S = \int_0^\pi 2\pi(2\sin\theta - \sin 2\theta)\cdot 2\sqrt{2}\;\sqrt{1 - \cos\theta}\; d\theta$$

$$= \int_0^\pi 8\sqrt{2}\;\pi\sin\theta(1 - \cos\theta)^{3/2}\; d\theta, \quad \text{with} \quad \sin 2\theta = 2\sin\theta\cos\theta$$

$$= \dfrac{16\sqrt{2}}{5}\;\pi(1 - \cos\theta)^{5/2}]_0^\pi = \dfrac{128\pi}{5} \approx 80.4 \quad \text{square units.}$$

11. The average value is

$$\dfrac{1}{2 - 0}\int_0^2 (x^2 - x + 1)\; dx = \tfrac{1}{2}[\tfrac{1}{3}x^3 - \tfrac{1}{2}x^2 + x]_0^2 = \tfrac{4}{3}\; .$$

12.   The planar region is shown at the right.
       We find that

$$\bar{x} = \frac{\int x \, dA}{\int dA} = \frac{\int_0^2 x(x + 2 - 2x) \, dx}{\int_0^2 (x + 2 - 2x) \, dx}$$

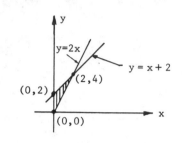

$$= \frac{\int_0^2 (2x - x^2) \, dx}{\int_0^2 (2 - x) \, dx} = \frac{x^2 - \frac{1}{3}x^3]_0^2}{2x - \frac{1}{2}x^2]_0^2} = \frac{2}{3}, \quad \text{and}$$

$$\bar{y} = \frac{\int y \, dA}{\int dA} = \frac{\int_0^2 y\left(\frac{y}{2}\right) \, dy + \int_2^4 y\left[\frac{y}{2} - (y - 2)\right] \, dy}{2}$$

$$= \frac{1}{2} \left[\left(\frac{1}{6}y^3\right)]_0^2 + \left(y^2 - \frac{1}{6}y^3\right)]_2^4\right] = \frac{1}{2}\left(\frac{4}{3} + 16 - \frac{32}{3} - 4 + \frac{4}{3}\right) = 2.$$

13.   A cross section of the half-filled pipe
       is shown at the right. Here
       $r = 4$ inches $= \frac{1}{3}$ foot. By the
       Theorem of Pythagoras,
       $\left(\frac{\ell}{2}\right)^2 + h^2 = r^2$ so that

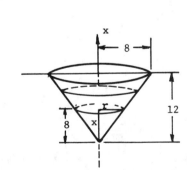

       $\ell = \frac{2}{3} \sqrt{1 - 9h^2}$.
       The force F is given by

$$F = \int wh \, dA = \int_0^{1/3} wh\ell \, dh = \frac{2w}{3} \int_0^{1/3} h\sqrt{1 - 9h^2} \, dh$$

$$= \left(\frac{2w}{3}\right)\left(\frac{2}{3}\right)\left(-\frac{1}{18}\right)(1 - 9h^2)^{3/2}]_0^{1/3} = \frac{2w}{81} \approx 1.54 \quad \text{pounds.}$$

14. $F = kx$ so $90 = k \cdot 1$ or $k = 90$. Thus, the work done is given

       by $W = \int F \, dx = \int_0^5 90x \, dx = 45x^2]_0^5 = 1125$ ft.-lb.

15.   The work done is $W = \int F \, ds$.
       The force is the weight of a typical
       volume of water. From the figure at
       the right,
       $\frac{r}{x} = \frac{8}{12}$ or $r = \frac{2}{3}x$.
       Thus, a typical volume element in a
       horizontal slice of the water is
       $dV = \pi r^2 \, dx = \frac{4\pi}{9} x^2 \, dx$.
       The distance this volume element is lifted is
       $12 - x + 2 = 14 - x$ feet. Thus,

$$W = \int_0^8 w \cdot \frac{4\pi}{9} x^2(14 - x) \, dx = \frac{4\pi w}{9}\left(\frac{14}{3}x^3 - \frac{1}{4}x^4\right)]_0^8 = \frac{4\pi w \cdot 8^4}{27}$$

       $\approx 18.96\pi$ ft.-tons (assuming $w = 62.5$).

CHAPTER 6 TRANSCENDENTAL FUNCTIONS

6.1 INVERSE FUNCTIONS.

$\boxed{\text{OBJECTIVE}}$: Use the derivative rule for inverse functions (Rule 11) to calculate the derivative of the inverse for a specified function.

1. If f and g are inverse functions on suitably restricted domains, then $g(f(x)) =$ _____ and $f(g(y)) =$ _____. That is, the composite of g and f or of f and g is the _____ mapping.

2. Given a function $y = f(x)$, to find a formula for the inverse function g, interchange the letters ____ and ____ in the original equation and solve for _____.

3. If f and g are inverse functions on suitably restricted domains, then $g'(a) = 1/f'(b)$ where a and b are related by _____ and _____.

4. Let $f(x) = -6x + 2$ and let g denote the inverse of f. We wish to calculate the derivative $g'(14)$. First, $-6x + 2 = 14$ implies x = _____. Thus, $f(-2) = 14$ so b = _____ and a = _____ in Problem 3. Then, $f'(x) =$ _____ so that $g'(14) = \dfrac{1}{f'(\underline{\quad})} =$ _____.

5. To calculate the inverse of $y = -6x + 2$, interchange the letters x and y obtaining _____. Solving the resultant equation for y yields _____, or $g(x) =$ _____ is the inverse function of $f(x) = -6x + 2$. Calculating the derivative $g'(14)$ directly from the formula for $g(x)$ gives $-1/6$ as before.

    Remark. The advantage of the derivative formula for the inverse function given by Rule 11 of the text is that it provides for the calculation of the derivative $g'(a)$ even though a formula for the inverse function g is not known.

6. Let g be the inverse of $f(x) = x^2 + 4x - 3$ for $x > -2$. To find $g'(-6)$, first set $f(x) = x^2 + 4x - 3$ equal to _____ and solve the quadratic equation yielding $x = -3$ or x = _____. We reject $x = -3$ because -3 is outside the allowable interval $x > -2$. Thus, b = _____ and

---

1. x, y, identity   2. x, y, y   3. $b = g(a)$, $a = f(b)$

4. -2, $b = -2$ and $a = 14$, -6, -2, $-\frac{1}{6}$ 5. $x = -6y + 2$, $y = -\frac{1}{6}(x - 2)$, $g(x) = -\frac{1}{6}(x - 2)$

$a = f(b) =$ _____ . $\frac{d}{dx} (x^2 + 4x - 3) =$ _____   so

$f'(-1) =$ _____ .   Thus, $g'(-6) = \dfrac{1}{f'(\underline{\quad})} =$ _____ .   Notice

that we did not need a formula for the inverse function g
itself.

7.   If $y = 4 - 7x$,   then the inverse function is $g(x) =$ _____ .

6-2   THE INVERSE TRIGONOMETRIC FUNCTIONS.

OBJECTIVE A :   Find the values of inverse trigonometric functions at
selected points without the use of tables or a
calculator.

8. $y = \sin^{-1} x$ is equivalent to _____ , where _____ $\leq x \leq$ _____
and _____ $\leq y \leq$ _____ .

9. $y = \tan^{-1} x$ is equivalent to _____ , where _____ $< x <$ _____
and _____ $< y <$ _____ .

10.   Let $y = \sin^{-1} \left(-\frac{\sqrt{2}}{2}\right)$;   then $\sin y =$ _____ ,   so $y =$ _____ .

That is, $\sin^{-1} \left(-\frac{\sqrt{2}}{2}\right) =$ _____ .

11.   If $\alpha = \tan^{-1} \left(-\frac{\sqrt{3}}{3}\right)$,   then $\tan \alpha =$ _____ ,   so $\alpha =$ _____ .

Hence, $\sin \alpha =$ _____ and $\cos \alpha =$ _____ .

OBJECTIVE B :   Simplify expressions involving inverse trigonometric
functions and trigonometric functions.

12.   To find $\sin\left(\cos^{-1} \left(-\frac{\sqrt{3}}{2}\right)\right)$,   let $y = \cos^{-1} \left(-\frac{\sqrt{3}}{2}\right)$.   Then,

$\cos y =$ _____ so $y =$ _____ .   Hence, $\sin y =$ _____ .
Alternatively since $\sin^2 y + \cos^2 y = 1$ holds,

$$\sin \left(\cos^{-1} \left(-\frac{\sqrt{3}}{2}\right)\right) = \sqrt{1 - \cos^2} \left(\cos^{-1} \left(-\frac{\sqrt{3}}{2}\right)\right)$$

$$= \left[1 - \cos^2\left(\cos^{-1}\left(-\frac{\sqrt{3}}{2}\right)\right)\right]^{1/2}$$

$$= \sqrt{1 - \underline{\quad}} = \underline{\quad} .$$

---

6. -6, -1, -1, -6, $2x + 4$, 2, -1, $\frac{1}{2}$ 7. $\frac{1}{7}(4 - x)$ 8. $x = \sin y$, -1, 1, $-\pi/2$, $\pi/2$

9. $x = \tan y$, $-\infty$, ∞, $-\frac{\pi}{2}$, $\frac{\pi}{2}$ 10. $-\frac{\sqrt{2}}{2}$, $-\frac{\pi}{4}$, $-\frac{\pi}{4}$

11. $-\frac{\sqrt{3}}{3}$, $-\frac{\pi}{6}$, $-\frac{1}{2}$, $\frac{\sqrt{3}}{2}$ 12. $-\frac{\sqrt{3}}{2}$, $\frac{5\pi}{6}$, $\frac{1}{2}$, $\frac{3}{4}$, $\frac{1}{2}$

6-3   THE DERIVATIVES OF THE INVERSE TRIGONOMETRIC FUNCTIONS: RELATED INTEGRALS.

OBJECTIVE A :   Differentiate functions whose expressions involve inverse trigonometric functions.

13. $\frac{d}{dx} \sin^{-1} u = $ _____ .

14. $\frac{d}{dx} \tan^{-1} u = $ _____ .

15. $\frac{d}{dx} \sec^{-1} u = $ _____ .

16. $\frac{d}{dx} \left(\sin^{-1} \frac{x}{5} \right)^2 = \left(2 \sin^{-1} \frac{x}{5} \right) \left(\frac{1}{\sqrt{1 - (x/5)^2}} \right) \frac{d}{dx} (\underline{\quad})$

$$= \left(2 \sin^{-1} \frac{x}{5} \right) \left(\frac{5}{\sqrt{25 - x^2}} \right) (\underline{\quad}) = \underline{\qquad} .$$

17. $\frac{d}{dx} \sec^{-1} \frac{1}{x} = \frac{1}{\frac{1}{|x|} \sqrt{\underline{\quad} - 1}} \cdot (\underline{\quad}) = \frac{-1}{\sqrt{\underline{\quad}}} .$

18. $\frac{d}{dx} \tan^{-1} \sqrt{x - 1} = \frac{1}{1 + (x - 1)} \frac{d}{dx} (\underline{\qquad}) = \underline{\qquad} .$

19.   Differentiating $\tan^{-1} \frac{x}{y} = \frac{1}{2}$ implicitly, we find

$$0 = \frac{d}{dx} \tan^{-1} \frac{x}{y} = \frac{1}{1 + (x/y)^2} \cdot \frac{d}{dx} (\underline{\quad}) = \frac{y^2}{x^2 + y^2} (\underline{\qquad})$$

$$= \frac{y - \underline{\quad}}{x^2 + y^2} ; \quad \text{thus} \quad \frac{dy}{dx} = \underline{\quad} .$$

OBJECTIVE B :   Evaluate definite integrals of functions whose anti-derivatives involve inverse trigonometric functions.

20. $\int_0^1 \frac{x\, dx}{\sqrt{1 - x^4}}.$ Let $u = x^2$, then $du = $ _____ so that

$x\, dx = $ _____ , and the indefinite integral becomes,

$$\int \frac{x\, dx}{\sqrt{1 - x^4}} = \frac{1}{2} \int \frac{du}{\underline{\quad}} = \underline{\qquad} + C. \quad \text{Thus,}$$

$$\int_0^1 \frac{x\, dx}{\sqrt{1 - x^4}} = \underline{\qquad} \Big]_0^1 = \frac{1}{2} \left(\sin^{-1} 1 - \underline{\quad} \right) = \underline{\quad} .$$

---

13. $\dfrac{du/dx}{\sqrt{1 - u^2}}$ 14. $\dfrac{du/dx}{1 + u^2}$ 15. $\dfrac{du/dx}{|u| \sqrt{u^2 - 1}}$ 16. $\frac{x}{5}, \frac{1}{5}, \dfrac{2}{\sqrt{25 - x^2}} \sin^{-1} \frac{x}{5}$

17. $\frac{1}{x^2}, -\frac{1}{x^2}, 1 - x^2$ 18. $\sqrt{x - 1}, \dfrac{1}{2x \sqrt{x - 1}}$

19. $\frac{x}{y}, \dfrac{y - x\frac{dy}{dx}}{y^2}, x\frac{dy}{dx}, \frac{y}{x}$ 20. $2x\, dx, \frac{1}{2} du, \sqrt{1 - u^2}, \frac{1}{2} \sin^{-1} u, \frac{1}{2} \sin^{-1} x^2, \sin^{-1} 0, \frac{\pi}{4}$

21. $\int_0^{\pi/2} \dfrac{\cos x \, dx}{1 + \sin^2 x}$.   Let $u = \sin x$,   then $du = $ _____,   and the indefinite integral becomes

$$\int \frac{\cos x \, dx}{1 + \sin^2 x} = \int \frac{du}{\underline{\hspace{2cm}}} = \underline{\hspace{2cm}} + C. \quad \text{Thus,}$$

$$\int_0^{\pi/2} \frac{\cos x \, dx}{1 + \sin^2 x} = \underline{\hspace{3cm}} \Big]_0^{\pi/2} = \tan^{-1} 1 - \underline{\hspace{2cm}}$$

$$= \underline{\hspace{1.5cm}} - 0 = \underline{\hspace{1.5cm}}.$$

$\boxed{\text{OBJECTIVE C}}$:   Use l'Hôpital's rule to find limits of indeterminate forms involving inverse trigonometric functions.

22. $\displaystyle\lim_{x \to 0} \dfrac{\sin^{-1} x}{\sin x}$ is of the form $0/0$,   so l'Hôpital's rule applies.

Thus, $\displaystyle\lim_{x \to 0} \dfrac{\sin^{-1} x}{\sin x} = \lim_{x \to 0} \dfrac{\underline{\hspace{1.5cm}}}{\cos x} = \dfrac{\underline{\hspace{1.5cm}}}{1} = \underline{\hspace{1.5cm}}.$

23. $\displaystyle\lim_{x \to +\infty} \dfrac{\frac{\pi}{2} - \tan^{-1} x}{1/x} = \lim_{t \to 0^+} \dfrac{\frac{\pi}{2} - \tan^{-1} \frac{1}{t}}{\underline{\hspace{1.5cm}}}$

$$= \lim_{t \to 0^+} \left(\underline{\hspace{1.5cm}} \right) \frac{d}{dt} \left(\frac{1}{t} \right) = \lim_{t \to 0^+} \left(- \frac{t^2}{t^2 + 1} \right) \left(\underline{\hspace{1.5cm}} \right)$$

$$= \lim_{t \to 0^+} \underline{\hspace{2cm}} = \underline{\hspace{2cm}}.$$

6-4   THE NATURAL LOGARITHM AND ITS DERIVATIVE.

$\boxed{\text{OBJECTIVE A}}$:   Use the trapezoidal rule to calculate natural logarithms of numbers between   0   and   10.

24. The natural logarithm is defined by $\ln x = $ _____   for x satisfying _____ .

25. By the First Fundamental Theorem of integral calculus, $\dfrac{d}{dx} \ln x = $ _____,   so the natural logarithm is a continuous function because it is _____ .

26. To find $\ln 2.5$,   we approximate $\int_1^{2.5} \frac{1}{x} \, dx$ by the trapezoidal rule. Let us take $n = 6$ subdivisions. Hence, $h = (2.5 - 1)/6 = $ _____.   Therefore, the subdivision points

---

21. $\cos x \, dx$, $1 + u^2$, $\tan^{-1} u$, $\tan^{-1}(\sin x)$, $\tan^{-1} 0$, $\frac{\pi}{4}$, $\frac{\pi}{4}$ 22. $\dfrac{1}{\sqrt{1 - x^2}}$, 1, 1

23. t, $-\dfrac{1}{1 + (1/t)^2}$, $-\dfrac{1}{t^2}$, $\dfrac{1}{t^2 + 1}$, 1   24. $\int_1^x \frac{1}{t} \, dt$, $x > 0$ 25. $\frac{1}{x}$, differentiable

are $x_0 = 1$, $x_1 = \frac{5}{4}$, $x_2 =$ _____, $x_3 =$ _____, $x_4 =$ _____, $x_5 =$ _____, and $x_6 = \frac{5}{2}$. Since $y = 1/x$, the trapezoidal approximation gives

$$T = \frac{h}{2}\left(\frac{1}{x_0} + \frac{2}{x_1} + \frac{2}{x_2} + \frac{2}{x_3} + \frac{2}{x_4} + \frac{2}{x_5} + \frac{1}{x_6}\right) = \frac{1}{8}\left(\underline{\hspace{5cm}}\right)$$

$$\approx \frac{\underline{\hspace{2cm}}}{8} \approx \underline{\hspace{2cm}}. \quad \text{Hence, } \ln 2.5 \approx \underline{\hspace{2cm}}.$$

27. Let us estimate the error in the approximation. For $f(x) = \frac{1}{x}$, the error is given by

$$E = \left|\frac{b - a}{12} f''(c) \cdot h^2\right| = \left|\frac{1.5}{12}(\underline{\hspace{1cm}}) \frac{1}{16}\right|$$

$$= \left|\frac{1}{\underline{\hspace{1cm}}}\right| < \underline{\hspace{2cm}} \quad \text{since } 1 < c < 2.5.$$

Therefore, because $\frac{1}{64} \approx 0.0156$ the trapezoidal approximation is accurate to at least one decimal place. (The actual error is, in fact, about 43×10^{-4}.)

OBJECTIVE B :   Differentiate functions whose expressions involve the natural logarithmic function.

28. $\frac{d}{dx} \ln\left(5 + 2x^3\right)^4 = \frac{1}{\left(5 + 2x^3\right)^4} \frac{d}{dx}\left(\underline{\hspace{3cm}}\right)$

$$= \frac{1}{\left(5 + 2x^3\right)^4}\left[\underline{\hspace{3cm}}\right]\frac{d}{dx}\left(5 + 2x^3\right)$$

$$= \frac{\overline{\hspace{2cm}}}{\left(5 + 2x^3\right)^4} = \underline{\hspace{3cm}}.$$

29. $\frac{d}{dx}\left[\ln(\sin x)\right]^2 = \underline{\hspace{4cm}} \frac{d}{dx}\ln(\sin x)$

$$= 2\ln(\sin x) \cdot \underline{\hspace{2cm}} \cdot \frac{d}{dx}(\sin x)$$

$$= 2\csc x \ln(\sin x) \cdot \underline{\hspace{1.5cm}} = \underline{\hspace{3cm}}.$$

30. $\frac{d}{dx} x^2 \ln \sqrt{x} = 2x \ln \sqrt{x} + \underline{\hspace{2cm}} \frac{d}{dx} \ln \sqrt{x}$

$$= 2x \ln \sqrt{x} + \underline{\hspace{3cm}} \frac{d}{dx} \sqrt{x}$$

$$= 2x \ln \sqrt{x} + \underline{\hspace{3cm}} = \frac{x}{2}\left(\underline{\hspace{3cm}}\right).$$

OBJECTIVE C :   Integrate functions whose antiderivatives involve the natural logarithm function.

31. $\int \frac{x \, dx}{x^2 + 4}$

Let $u = x^2 + 4$. Then $du =$ _____ so $x \, dx =$ _____. Thus the integral becomes

---

26. $\frac{1}{4}$, $\frac{3}{2}$, $\frac{7}{4}$, 2, $\frac{9}{4}$, $\left(1 + \frac{8}{5} + \frac{4}{3} + \frac{8}{7} + 1 + \frac{8}{9} + \frac{2}{5}\right)$, 7.3651, .9206, .9206   27. $-2/c^3$, $64c^3$, $1/64$

28. $\left(5 + 2x^3\right)^4$, $4\left(5 + 2x^3\right)^3$, $24x^2\left(5 + 2x^3\right)^3$, $\frac{24x^2}{5 + 2x^3}$ 29. $2\ln(\sin x)$, $\frac{1}{\sin x}$, $\cos x$, $2\cot x \ln(\sin x)$

30. x^2, $x^2 \cdot \frac{1}{\sqrt{x}}$, $x^2 \cdot \frac{1}{\sqrt{x}} \cdot \frac{1}{2\sqrt{x}}$, $2\ln x + 1$ or, $4\ln\sqrt{x} + 1$

$$\int \frac{x\ dx}{x^2 + 4} = \int \frac{du}{\underline{\hspace{1cm}}} = \underline{\hspace{2.5cm}} + C = \underline{\hspace{3cm}}.$$

32. $$\int \frac{3x + 1}{x}\ dx = \int (3 + \underline{\hspace{1cm}})\ dx = \int 3\ dx + \underline{\hspace{2cm}}$$

$$= 3x + \underline{\hspace{2.5cm}} + C.$$

33. $$\int \frac{dx}{x\ \ln\ \sqrt{x}}$$

Let $u = \ln\ \sqrt{x}$.   Then $\frac{du}{dx} = \underline{\hspace{1.5cm}}\ \frac{d}{dx}(\underline{\hspace{1cm}}) = \underline{\hspace{2cm}}.$

Hence, $2\ du = \underline{\hspace{2cm}}\ dx.$ Thus the integral becomes

$$\int \frac{dx}{x\ \ln\ \sqrt{x}} = \int \frac{2\ du}{\underline{\hspace{1cm}}} = \underline{\hspace{2.5cm}} + C = \underline{\hspace{3cm}}.$$

OBJECTIVE D :   Use l'Hôpital's rule to find limits of indeterminate forms involving the natural logarithm function.

34. $$\lim_{x \to \infty} \frac{\ln\left(\frac{1 + x}{x}\right)}{1/x} = \lim_{x \to \infty} \frac{\ln\left(1 + \frac{1}{x}\right)}{1/x} = \lim_{t \to 0} \frac{\ln\ (1 + t)}{\underline{\hspace{1.5cm}}}$$

$$= \lim_{t \to 0} \frac{\underline{\hspace{1.5cm}}}{1} = \frac{1}{1 + \underline{\hspace{1cm}}} = \underline{\hspace{1.5cm}}.$$

35. $$\lim_{x \to 0^+} x^2\ \ln\ x$$

As $x \to 0^+$, $x^2 \to 0$ and $\ln x \to \underline{\hspace{1.5cm}}$ so this is an indeterminate form $0 \cdot \infty$.   Writing the limit as

$$\lim_{x \to 0^+} x^2\ \ln\ x = \lim_{x \to 0^+} \frac{\ln\ x}{\underline{\hspace{1cm}}},$$ we see this form is now of the type

∞/∞.   Therefore l'Hôpital's rule applies:

$$\lim_{x \to 0^+} \frac{\ln\ x}{1/x^2} = \lim_{x \to 0^+} \underline{\hspace{2cm}} = \lim_{x \to 0^+} \underline{\hspace{2cm}} = \underline{\hspace{2cm}}.$$ Hence,

$$\lim_{x \to 0^+} x^2\ \ln\ x = \underline{\hspace{2cm}}.$$

36. $$\lim_{\theta \to \frac{\pi}{2}^-} \frac{\ln\ (\tan\ \theta)}{\sec\ \theta}$$

As $\theta \to \frac{\pi}{2}^-$, $\tan\ \theta \to \infty$ and $\sec\ \theta \to \underline{\hspace{1.5cm}}.$ Hence $\ln\ (\tan\ \theta) \to \underline{\hspace{1.5cm}}$ and this is an indeterminate form of type $\underline{\hspace{2cm}}$.   Therefore, l'Hôpital's rule applies:

$$\lim_{\theta \to \frac{\pi}{2}^-} \frac{\ln\ (\tan\ \theta)}{\sec\ \theta} = \lim_{\theta \to \frac{\pi}{2}^-} \frac{\underline{\hspace{1.5cm}}}{\sec\ \theta\ \tan\ \theta} = \lim_{\theta \to \frac{\pi}{2}^-} \frac{\underline{\hspace{1.5cm}}}{\sin^2\ \theta} = \underline{\hspace{2cm}}.$$

---

31. $2x\ dx$, $\frac{1}{2}\ du$, $2u$, $\frac{1}{2}\ \ln\ |u|$, $\frac{1}{2}\ \ln\ (x^2 + 4) + C$ 32. $\frac{1}{x}$, $\int \frac{dx}{x}$, $\ln\ |x|$

33. $\frac{1}{\sqrt{x}}$, $\sqrt{x}$, $\frac{1}{2x}$, $\frac{1}{x}$, u, $2\ \ln\ |u|$, $2\ \ln\ (\ln\ \sqrt{x}) + C$ 34. t, $1/(1 + t)$, 0, 1

35. $-\infty$, $1/x^2$, $\frac{1/x}{-2/x^3}$, $-\frac{1}{2}\ x^2$, 0, 0

36. ∞, ∞, ∞/∞, $\sec^2\theta/\tan\ \theta$, $\cos\ \theta$, 0

6-5   PROPERTIES OF NATURAL LOGARITHMS.   THE GRAPH OF $y = \ln x$.

OBJECTIVE A : Use the three properties of the natural lograithm to rewrite a logarithmic expression as a sum, difference, or multiple of logarithms.

37. $\ln ax =$ _____ for $a > 0$ and $x > 0$.

38. $\ln \frac{x}{a} =$ _____ for $a > 0$ and $x > 0$.

39. $\ln x^n =$ _____ for $x > 0$ and n rational.

40. $\ln \sqrt[3]{\frac{x^2}{a^4}} = \ln\left(\frac{x^2}{a^4}\right)^{-} = ($____$) \ln \left(\frac{x^2}{a^4}\right) = \frac{1}{3} \left(\ln x^2 -$_____$\right)$

   $= \frac{2}{3} \ln x -$_____.

41. $\ln (b^3 \cdot \sqrt{x}) = \ln b^3 +$_____$= 3$_____$+ \ln x^{1/2} =$_____.

42. $\ln (x^2 + 2x + 1) = \ln (x + 1)^{-} =$_____ for $x > -1$.

OBJECTIVE B : Summarize the characteristics of the graph of $y = \ln x$,   and graph functions involving the natural logarithm.

43.   The domain of $y = \ln x$ is the set _____, and its range is the set _____.

44.   The graph of $y = \ln x$ is increasing _____. It is concave downward _____.

45.   Since $y = \ln x$ is differentiable for $x > 0$,   it is a _____ function of x.

46. $\lim\limits_{x \to \infty} \ln x =$ _____   and $\lim\limits_{x \to 0^+} \ln x -$_____.

47.   Consider the curve $y = x - \ln x$.   The derivative $\frac{dy}{dx} =$_____ so that $\frac{dy}{dx} = 0$ implies $\frac{1}{x} =$_____   or $x =$_____.   Notice that the domain of y is the set _____.   The second derivative $\frac{d^2y}{dx^2} =$_____   is always positive.   Therefore, the critical point $x = 1$ gives a relative _____ value of $y(1) =$_____.   As $x \to 0$, $y \to$_____.   To examine the curve as $x \to \infty$,   notice that

---

37. $\ln a + \ln x$ 38. $\ln x - \ln a$ 39. $n \ln x$ 40. $\frac{1}{3}$, $\frac{1}{3}$, $\ln a^4$, $\frac{4}{3} \ln a$

41. $\ln \sqrt{x}$, $\ln b$, $3 \ln b + \frac{1}{2} \ln x$ 42. 2, $2 \ln (x + 1)$ 43. $x > 0$, $-\infty < y < +\infty$

44. for all x in the domain, for all $x > 0$ 45. continuous   46. $+\infty$, $-\infty$

$$\lim_{x\to\infty} \left(\frac{x - \ln x}{x}\right) = 1 - \lim_{x\to\infty} \frac{\ln x}{x} = 1 - \lim_{x\to\infty} \underline{\hspace{1.5cm}} = \underline{\hspace{1.5cm}} \quad \text{by}$$

l'Hôpital's rule. Hence, for large values of x, the ratio $\frac{x - \ln x}{x}$ is approximately

_____ or x - ln x ≈ _____ .
Sketch a graph of y at the right.

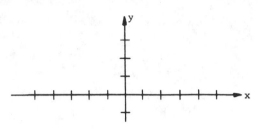

OBJECTIVE C :   Use the method of logarithmic differentiation to calculate derivatives.

48. Find $\frac{dy}{dx}$ if $y = x^{\tan x}$,   x > 0.

Solution.
ln y = ln(x^{tan x}) = _____ so that

$\frac{1}{y} \frac{dy}{dx}$ = sec²x ln x + _____ ,   or

$\frac{dy}{dx}$ = x^{tan x} (_____).

49. Find $\frac{dy}{dx}$ if $y = \sqrt{\frac{1 - x}{1 + x}}$,   -1 < x < 1.

Solution.
$\ln y = \ln \sqrt{\frac{1 - x}{1 + x}} = \frac{1}{2} \ln (1 - x) - \underline{\hspace{3cm}}$ so that

$\frac{1}{y} \frac{dy}{dx} = - \frac{1}{2(1 - x)} - \underline{\hspace{1.5cm}} = \frac{-(1 + x) - (\underline{\hspace{0.8cm}})}{2(1 - x)(1 + x)}$

$\frac{dy}{dx} = -y(1 - x)^{-1} \cdot \underline{\hspace{1.5cm}} = \underline{\hspace{3cm}}$.

50. Find $\frac{dy}{dx}$ if $y = \left(x^r\right)^x$,   x > 0.
Solution.

$\ln y = \ln \left(x^r\right)^x = \underline{\hspace{2.5cm}} = \underline{\hspace{2.5cm}}$.
$\frac{1}{y} \frac{dy}{dx} = r \ln x + \underline{\hspace{1.5cm}} = r(\underline{\hspace{2cm}})$ so that

$\frac{dy}{dx} = \underline{\hspace{3cm}}$.

---

47. $1 - \frac{1}{x}$, 1, 1, x > 0, $\frac{1}{x^2}$, minimum, 1, +∞, $\frac{1}{x}$, 1, 1, x

48. tan x · ln x, tan x · $\frac{1}{x}$, sec² x ln x + $\frac{1}{x}$ tan x

49. $\frac{1}{2}$ ln (1 + x), $\frac{1}{2(1 + x)}$, 1 - x, (1 + x)⁻¹,

   -(1 - x)^{-1/2}(1 + x)^{-3/2}

50. x ln x^r, rx ln x, rx · $\frac{1}{x}$, ln x + 1, r(x^r)^x (ln x + 1)

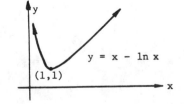

6-6 THE EXPONENTIAL FUNCTION e^x.

OBJECTIVE A : Use the equivalent equations $y = e^x$ and $x = \ln y$ to simplify logarithms of exponentials, and exponentials of logarithms.

51. The equation $y = e^{\ln x}$ is equivalent to $\ln y = $ _____.
Since the logarithm is one - one, the last equation is equivalent to _____; that is, $e^{\ln x} = $ _____. In other words, the exponential "undoes" the natural logarithm.

52. The equation $y = \ln (e^x)$ is equivalent to $e^y = $ _____.
Since the exponential function is one - one, the last equation is equivalent to _____; that is, $\ln (e^x) = $ _____.

53. $e^{-2 \ln (x + 1)} = e^{\ln \underline{\qquad}} = $ _____ by Problem 51.

54. $\ln (\sqrt{x}e^{2-x}) = \ln \sqrt{x} + $ _____ $= $ _____.

OBJECTIVE B : Differentiate functions whose expressions involve exponential functions.

55. $\dfrac{d}{dx} e^{\sqrt{1 - x^2}} = e^{\sqrt{1 - x^2}} \cdot \dfrac{d}{dx} ($_____$)$

$\qquad = e^{\sqrt{1 - x^2}} \cdot \dfrac{1}{2\sqrt{1 - x^2}} \cdot \dfrac{d}{dx} ($_____$)$

$\qquad = $ _____.

56. $\dfrac{d}{dx} e^{x \ln x} = $ _____ $\cdot \dfrac{d}{dx} (x \ln x) = e^{x \ln x} ($_____$)$.

57. $\dfrac{d}{dx} \ln (e^x + 1) = \dfrac{1}{e^x + 1} \cdot \dfrac{d}{dx} ($_____$) = $ _____.

58. $\dfrac{d}{dx} \sin \sqrt{e^x} = \cos \sqrt{e^x} \cdot \dfrac{d}{dx} ($_____$)$

$\qquad = \cos \sqrt{e^x} \cdot \left(\dfrac{1}{2\sqrt{e^x}}\right) \cdot \dfrac{d}{dx} ($_____$)$

$\qquad = $ _____.

OBJECTIVE C : Integrate functions whose antiderivatives involve exponential functions.

59. $\displaystyle\int x^2 e^{-x^3} \, dx$

Let $u = -x^3$. Then $du = $ _____ so that $x^2 \, dx = $ _____.

---

51. $\ln x$, $y = x$, x

52. e^x, $y = x$, x

53. $(x + 1)^{-2}$, $(x + 1)^{-2}$

54. $\ln (e^{2-x})$, $\ln \sqrt{x} + (2 - x)$

55. $\sqrt{1 - x^2}$, $1 - x^2$, $\dfrac{-x}{\sqrt{1 - x^2}} e^{\sqrt{1 - x^2}}$

56. $e^{x \ln x}$, $\ln x + 1$

57. $e^x + 1$, $\dfrac{e^x}{e^x + 1}$

58. $\sqrt{e^x}$, e^x, $\frac{1}{2} \sqrt{e^x} \cos \sqrt{e^x}$

Thus the integral becomes,

$$\int x^2 e^{-x^3}\, dx = \int \underline{\hspace{2cm}}\, du = \underline{\hspace{1.5cm}} + C = \underline{\hspace{3cm}}.$$

60. $\int \left(e^x - 2\right)^4 e^x\, dx$

Let $u = e^x - 2$. Then $du = \underline{\hspace{2cm}}$ and the integral becomes,

$$\int \left(e^x - 2\right)^4 e^x\, dx = \int \underline{\hspace{1.5cm}}\, du = \underline{\hspace{1.5cm}} + C = \underline{\hspace{3cm}}.$$

61. $\int \dfrac{e^x}{1 + e^{2x}}\, dx$

Let $u = e^x$. Then $du = \underline{\hspace{1.5cm}}$ and the integral becomes,

$$\int \frac{e^x\, dx}{1 + e^{2x}} = \int \frac{du}{\underline{\hspace{1cm}}} = \underline{\hspace{2cm}} + C = \underline{\hspace{3cm}}.$$

OBJECTIVE D : Use l'Hôpital's rule to find limits of indeterminate forms involving exponential functions.

62. $\displaystyle\lim_{x\to\infty} xe^{-x} = \lim_{x\to\infty} \frac{x}{e^x}$ is an indeterminate form of type $\underline{\hspace{1.5cm}}$ so l'Hôpital's rule applies. Thus, $\displaystyle\lim_{x\to\infty} \frac{x}{e^x} = \lim_{x\to\infty} \underline{\hspace{1.5cm}} = \underline{\hspace{1cm}}.$

63. $\displaystyle\lim_{x\to 0} \frac{e^{2x} - 1}{x}$ is an indeterminate form of type $\underline{\hspace{1cm}}$ so l'Hôpital's rule applies. Thus,

$$\lim_{x\to 0} \frac{e^{2x} - 1}{x} = \lim_{x\to 0} \underline{\hspace{1cm}} = \frac{\underline{\hspace{1cm}}}{1} = \underline{\hspace{1.5cm}}.$$

OBJECTIVE E : Graph functions whose expressions contain exponential functions.

64. Consider the function $y = xe^x$. The domain of this function is the set $\underline{\hspace{2cm}}$. To find the critical points: $dy/dx = 0$ implies $e^x + \underline{\hspace{2cm}} = 0$, or $\underline{\hspace{1.5cm}} = 0$. Hence,

$x = \underline{\hspace{1.5cm}}$ is the only critical point. $\dfrac{d^2y}{dx^2} = e^x\,(\underline{\hspace{1.5cm}})$

which is positive when $x = -1$; thus $x = -1$ gives a relative

$\underline{\hspace{3cm}}$ value of $y = \underline{\hspace{2.5cm}}$. $\dfrac{d^2y}{dx^2} = 0$ gives a

point of inflection when $x = \underline{\hspace{1.5cm}}$. For $x < -2$

---

59. $-3x^2\, dx$, $-\frac{1}{3}\, du$, $-\frac{1}{3}\, e^u$, $-\frac{1}{3}\, e^u$, $-\frac{1}{3}\, e^{-x^3} + C$ 60. $e^x\, dx$, u^4, $\frac{1}{5}\, u^5$, $\frac{1}{5}\left(e^x - 2\right)^5 + C$

61. $e^x\, dx$, $1 + u^2$, $\tan^{-1} u$, $\tan^{-1}\left(e^x\right) + C$ 62. ∞/∞, $\frac{1}{e^x}$, 0

63. $0/0$, $2e^{2x}$, 2

the curve is concave _____, and
for x > -2 it is concave _____.
By l'Hôpital's rule,

$$\lim_{x \to -\infty} x e^x = \lim_{x \to -\infty} \frac{x}{e^{-x}} = \lim_{x \to -\infty} \frac{1}{\underline{\quad}} =$$

_____. Hence, the x-axis is a
horizontal asymptote. As x → +∞,
y → _____. Sketch the graph
in the coordinate system at the right.

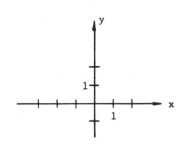

6-7 THE FUNCTIONS a^x AND a^u.

65. The function y = a^x is defined by a^x = _____ and
it is well-defined whenever _____.

66. The definition in Problem 65 is equivalent to saying that
ln a^x = _____.

67. The derivative of a^u, where u is a differentiable function
of x, is given by $\frac{d}{dx} a^u$ = _____.

OBJECTIVE A : Differentiate functions whose expressions involve an
exponential function a^u, where u is a differentiable function of x.

68. $\frac{d}{dx} 2^{\sec x}$ = _____ · $\frac{d}{dx} \sec x$ = _____.

69. $\frac{d}{dx} x^2 3^x$ = 2x · _____ + x^2 · _____ = 3^x (_____).

OBJECTIVE B : Integrate functions whose antiderivatives involve an
exponential function a^u.

70. $\int \frac{dx}{2^x} = \int 2^{-x} dx$

Let u = -x so that du = _____, and the integral becomes

$\int 2^{-x} dx = \int$ _____ du = _____ + C = _____.

---

64. -∞ < x < ∞, xe^x, e^x(1 + x), -1, x + 2, minimum,
-1/e, -2, downward, upward, -e^{-x}, 0, +∞

65. e^{x ln a}, a > 0   66. x ln a

67. $a^u \cdot \frac{du}{dx} \cdot \ln a$ 68. $2^{\sec x} \cdot \ln 2$, (sec x tan x) $2^{\sec x} \cdot \ln 2$

69. 3^x, 3^x ln 3, 2x + x^2 ln 3

70. -dx, -2^u, $-\frac{1}{\ln 2} 2^u$, $-\frac{1}{2^x \ln 2} + C$

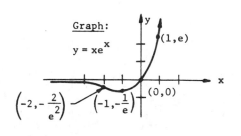

Graph:
y = xe^x
(1, e)
$\left(-2, -\frac{2}{e^2}\right)$ $\left(-1, -\frac{1}{e}\right)$ (0, 0)

71. $\int_1^2 x \; 10^{x^2-1} \; dx$

Let $u = x^2 - 1$. Then $du = $ _____ so that
$x \; dx = $ _____.

$\int x \; 10^{x^2-1} \; dx = \int$ _____ $du = $ _____ $+ \; C.$ Hence,

$\int_1^2 x \; 10^{x^2-1} \; dx = $ _____ $\bigg]_1^2 = \dfrac{1}{2 \ln 10} ($ _____ $)$

$= $ _____ .

[OBJECTIVE C]:   Graph functions whose expressions involve an
exponential function a^u.

72. Consider the curve $y = (0.2)^{x+1} + 2$.
$\dfrac{dy}{dx} = $ _____ $\dfrac{d}{dx} (x + 1) = $ _____ .
Therefore, dy/dx is of constant _____ sign, since
$\ln(.2) < 0$, and the curve is everywhere _____ .
Calculation of the second derivative gives $\dfrac{d^2y}{dx^2} = $ _____
which is of constant _____ sign; hence, the curve is
everywhere concave _____ .
As $x \to \infty$, $(.2)^{x+1} \to$ _____
so $y \to$ _____ . As $x \to -\infty$,
$(.2)^{x+1} \to$ _____ so $y \to$ _____ .
Finally, the points $(-1,$ _____ $)$,
$(0,$ _____ $)$, $(1,$ _____ $)$ lie on
the curve. You can now sketch the
graph of the curve in the
coordinate system provided at the
right.

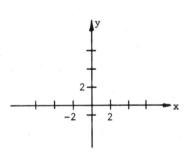

[OBJECTIVE D]:   Find limits of indeterminate forms involving exponential functions $[f(x)]^{g(x)}$.

73. $\lim\limits_{x \to 1} \dfrac{2^x - 2}{x - 1}$ is an indeterminate form of type _____ so

l'Hôpital's rule applies. Thus,

$\lim\limits_{x \to 1} \dfrac{2^x - 2}{x - 1} = \lim\limits_{x \to 1} \dfrac{\rule{2cm}{0.4pt}}{1} = $ _____ .

---

71. $2x \; dx$, $\frac{1}{2} \; du$, $\frac{1}{2} \; 10^u$, $\frac{1}{2 \ln 10} \; 10^u$, $\frac{1}{2 \ln 10} \; 10^{x^2-1}$, $10^3 - 1$, $\frac{999}{2 \ln 10}$

72. $(0.2)^{x+1} \ln(.2)$, $(0.2)^{x+1} \ln(.2)$, negative,
    decreasing, $(0.2)^{x+1}[\ln(.2)]^2$, positive,
    upward, 0, 2, $+\infty$, $+\infty$, 3, 2.2, 2.04

73. $0/0$, $2^x \ln 2$, $2 \ln 2$

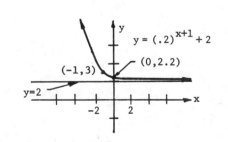

74. $\lim\limits_{x\to 0^+} x^{\sin x}$.   Let $y = x^{\sin x}$.   Then $\ln y = $ _____.

Thus $\ln y = \dfrac{\ln x}{\underline{\hspace{1cm}}}$.   Then $\lim\limits_{x\to 0^+} \dfrac{\ln x}{\csc x}$ is an indeterminate form

of type _____ and l'Hôpital's rule applies.   Hence,

$$\lim\limits_{x\to 0^+} \frac{\ln x}{\csc x} = \lim\limits_{x\to 0^+} \frac{1/x}{-\csc x \, \operatorname{ctn} x} = \lim\limits_{x\to 0^+} \frac{\overline{\hspace{1cm}}}{-x \cos x}$$

$$= \lim\limits_{x\to 0^+} \frac{\overline{\hspace{2cm}}}{x \sin x - \cos x} = \underline{\hspace{1.5cm}}.$$

Since $\ln y \to 0$ as $x \to 0^+$, $y \to$ _____.   Thus the required limit is _____.

6-8   THE FUNCTIONS $y = \text{LOG}_a u$: RATES OF GROWTH.

$\boxed{\text{OBJECTIVE A}}$:   Use the definition of $\log_a u$ to evaluate simple expressions.

75.   If $a > 0$ and $a \neq 1$,   then $y = \log_a u$ is defined and equivalent to $a^y = $ _____ whenever u is positive.   If a or u is negative, $\log_a u$ is _____.

76.   In terms of natural logarithms, $\log_a u = $ _____.

77. $\log_8 4$
If $y = \log_8 4$,   then $8^y = $ _____ or $2^{\underline{\hspace{0.5cm}}} = 2^2$.   Thus, $3y = $ _____ or $y = $ _____.   Therefore, $\log_8 4 = $ _____.

78. $\log_{.75} \dfrac{27}{64}$

If $y = \log_{.75} \dfrac{27}{64}$,   then $\dfrac{27}{64} = $ _____.   Now, $\left(\dfrac{3}{4}\right)^3 = $ _____

so that $\log_{.75} \dfrac{27}{64} = $ _____.

$\boxed{\text{OBJECTIVE B}}$:   Solve exponential and logarithmic equations.

79. $5^{x^2} = 7^x$
Taking the natural logarithm of both sides gives x^2 (_____) $= x \ln 7$ or x (_____) $= 0$.   Hence, $x = 0$ and $x = $ _____ both solve the equation.

80. $\log_7 (x^2 - 6x) = 1$ is equivalent to $x^2 - 6x = $ _____ or, $0 = $ _____ $= (x - 7)($ _____ $)$.   Therefore, $x = $ _____ and $x = $ _____ solve the logarithmic equation.

---

74. $\sin x \cdot \ln x$, $\csc x$, ∞/∞, $\sin^2 x$, $2 \sin x \cos x$, 0, 1, 1 75. u, defined

76. $\dfrac{\ln u}{\ln a}$ 77. 4, $3y$, 2, $\dfrac{2}{3}$, $\dfrac{2}{3}$ 78. $\left(\dfrac{3}{4}\right)^y$, $\dfrac{27}{64}$, 3

79. $\ln 5$, $x \ln 5 - \ln 7$, $\dfrac{\ln 7}{\ln 5}$ 80. 7^1, $x^2 - 6x - 7$, $x + 1$, -1, 7

81. $5^{\log_5 2}$ = _____ because $y = 5^u$ and $y = \log_5 u$ are _____ functions of each other.

82. In general, $a^{\log_a u}$ = _____ and $\log_a a^x$ = _____.

[OBJECTIVE C]: Differentiate functions involving a logarithmic function $\log_a u$.

83. $\dfrac{d}{dx} \log_a u = \dfrac{d}{dx} \dfrac{}{\ln a}$ = _____ $\cdot \dfrac{du}{dx}$.

84. $\dfrac{d}{dx} \log_{10} (x^2 - e^x) = \dfrac{1}{\rule{1cm}{0.4pt}} \dfrac{d}{dx} (x^2 - e^x)$ = _____.

85. Find $\dfrac{dy}{dx}$ if $y = \left(1 + \sqrt{x}\right)^{\log_2 x}$

Solution.

$\ln y = \ln \left(1 + \sqrt{x}\right)^{\log_2 x}$ = _____.

$\dfrac{1}{y} \dfrac{dy}{dx}$ = _____ $\ln (1 + \sqrt{x}) + \log_2 x \cdot \dfrac{1}{1 + \sqrt{x}} \cdot$ _____

$\dfrac{dy}{dx} = \left(1 + \sqrt{x}\right)^{\log_2 x} [\underline{\hspace{6cm}}]$.

[OBJECTIVE D]: Given two exponential, polynomial, or logarithmic functions, determine whether one is growing faster than the other, or if they are growing at the same rate.

86. Compare the growths of the functions $y = x^2$ and $y = 2^x$. Solution. We just consider the ratio $\dfrac{x^2}{2^x}$ as $x \rightarrow$ _____.

Thus, $\displaystyle\lim_{x \to \infty} \dfrac{x^2}{2^x} = \lim_{x \to \infty} \dfrac{2x}{\rule{1.5cm}{0.4pt}} = \lim_{x \to \infty} \dfrac{2}{\rule{1.5cm}{0.4pt}}$ = _____.

6-9   APPLICATIONS OF EXPONENTIAL AND LOGARITHMIC FUNCTIONS.

[OBJECTIVE]: Solve exponential growth and decay problems: $\dfrac{dx}{dt} = kx$ with $x = x_0$ when $t = 0$.

87. The <u>half-life</u> of a radioactive substance is the length of time it takes for _____ of a given amount of the substance to disintegrate through radiation. The half-life of the carbon isotope C^{14} is about 5700 years.

88. Assume that the amount x of C^{14} present in a dead organism decays exponentially from the time of death. Then, $x = x_0 e^{kt}$,

---

81. 2, inverse   82. u, x   83. $\ln u$, $\dfrac{1}{u \ln a}$ 84. $(x^2 - e^x) \ln 10$, $\dfrac{2x - e^x}{(x^2 - e^x) \ln 10}$

85. $\log_2 x \cdot \ln (1 + \sqrt{x})$, $\dfrac{1}{x \ln 2}$, $\dfrac{1}{2\sqrt{x}}$, $\dfrac{1}{x \ln 2} \ln (1 + \sqrt{x}) + \dfrac{\ln x}{2 \ln 2 \cdot \sqrt{x}(1 + \sqrt{x})}$

86. ∞, $2^x \ln 2$, $2^x (\ln 2)^2$, 0, slower   87. half

where x_0 is the original amount present. To find the constant k in the case of carbon C^{14},

$$\left(\frac{1}{2}\right) x_0 = x_0\ e^{\underline{\quad} k} \quad \text{or} \quad \ln\frac{1}{2} = \underline{\hspace{3cm}}.$$

Thus, $k = \dfrac{-\ln 2}{\underline{\quad}} \approx -1.22 \times 10^{-4}$.

Suppose we want to determine the amount of C^{14} present after 10,000 years. Then, the percentage is given by the ratio

$$\frac{x_0\ e^{k \cdot 10^4}}{x_0} \approx e^{\underline{\hspace{1.5cm}}} \approx 0.2964,$$

or approximately 29.64 percent of the amount of C^{14} remains after 10,000 years.

89.   The "1470" skull found in Kenya by Richard Leakey is reputed to be 2,500,000 years old. The percentage of C^{14} remaining is given by

$$\frac{x_0\ e^{k \cdot 2.5 \times 10^6}}{x_0} \approx e^{\underline{\hspace{1cm}}}.$$

However, if $x < -21$ then $e^x < 10^{-9}$ so the percentage of C^{14} left in the skull would be negligible. The current reliable limit for C^{14} dating is about 40,000 years, so another method for dating the skull had to be found.

---

88. 5700, 5700k, 5700, $-\dfrac{\ln 2}{57} \times 10^2$ 89. -304

CHAPTER 6   SELF-TEST

In Problems 1-10 calculate the derivative $\frac{dy}{dx}$.

1. $y = \tan\left(\cos\frac{2}{x}\right)$

2. $y = x^2 e^{-1/x}$

3. $y = x(\ln x)^2$

4. $y = \cos^{-1}\left(\frac{1 - x^2}{1 + x^2}\right), \quad x > 0$

5. $y = 5^x \log_5 x$

6. $y = 2^{\sin^{-1} x}$

7. $y = 2x\tan^{-1} 2x - \ln\sqrt{1 + 4x^2}$

8. $y = \frac{\sec 3\sqrt{x}}{\sqrt{x}}$

9. $y = (x)^{\sqrt{x}}, \quad x > 0$

10. $e^y + e^x = e^{x+y}$

In Problems 11-17 calculate the indicated integrals.

11. $\int \sec^2 (3x + 5)\, dx$

12. $\int_{-5/4}^{5/4} \frac{dx}{25 + 16x^2}$

13. $\int \left(1 - e^{-x}\right)^2 dx$

14. $\int \frac{1 - 2\cos x}{\sin^2 x}\, dx$

15. $\int \frac{(\ln 2x)^3}{x}\, dx$

16. $\int \frac{e^{-x}}{\sqrt{1 - e^{-2x}}}\, dx$

17. $\int_1^e \frac{2^{\ln x}}{3x}\, dx$

In Problems 18-21 evaluate the limits using l'Hôpital's rule.

18. $\lim\limits_{x \to 0} \csc x \sin^{-1} x$

19. $\lim\limits_{x \to 1} x^{1/(1-x)}$

20. $\lim\limits_{x \to 0^+} \frac{e^x - \cos x}{x \sin x}$

21. $\lim\limits_{x \to 0} \frac{4^x - 2^x}{x}$

22. Determine the following values.

(a) $\sin^{-1}\left(-\frac{\sqrt{3}}{2}\right)$

(b) $\tan^{-1}\frac{\sqrt{3}}{3}$

(c) $\tan^{-1}\left(\cos\frac{\pi}{2}\right)$

(d) $\cos^{-1}\left(\sin\frac{\pi}{6}\right)$

23. Simplify the expression $\dfrac{\log_3 243}{\log_2 \sqrt[4]{64} + \log_8 8^{-10}}$

24. Let $f(x) = \log_a x$, $f(5) = 1.46$, $f(2) = 0.63$, $f(7) = 1.77$. Use the properties of logarithms to find,
(a) $f(10)$ (b) $f(49)$ (c) $f\left(\frac{5}{7}\right)$ (d) $f(1.4)$

25. Solve the following equations.
(a) $3^{-8x+6} = 27^{-x-8}$ (b) $\log_5 (5x - 1) = -2$
(c) $e^x = 10^{x+1}$

26. Graph the curve $y = \dfrac{2}{1 + 3e^{-2x}}$.

27. Sketch the graph of $y = \dfrac{\ln x}{x^2}$.

28. Suppose that the number of bacteria in a yeast culture grows at a rate proportional to the number present. If the population of a colony of yeast bacteria doubles in one hour, find the number of bacteria present at the end of 3.5 hours.

SOLUTIONS TO CHAPTER 6 SELF-TEST

1. $\dfrac{dy}{dx} = \sec^2\left(\cos\dfrac{2}{x}\right) \cdot \left(-\sin\dfrac{2}{x}\right) \cdot \left(-\dfrac{2}{x^2}\right)$

2. $\dfrac{dy}{dx} = 2xe^{-1/x} + x^2 e^{-1/x} \cdot \left(\dfrac{1}{x^2}\right) = e^{-1/x}(2x + 1)$

3. $\dfrac{dy}{dx} = (\ln x)^2 + 2x \ln x \cdot \dfrac{1}{x} = \ln x\,(2 + \ln x)$

4. $\dfrac{dy}{dx} = \dfrac{-1}{\sqrt{1 - \left(\dfrac{1 - x^2}{1 + x^2}\right)^2}} \cdot \left[\dfrac{(1 + x^2)(-2x) - (1 - x^2)(2x)}{(1 + x^2)^2}\right]$

$= \dfrac{-(1 + x^2)}{\sqrt{(1 + x^2)^2 - (1 - x^2)^2}} \cdot \dfrac{-4x}{(1 + x^2)^2} = \dfrac{4x}{(1 + x^2)\sqrt{4x^2}}$

$= \dfrac{2x}{(1 + x^2)|x|} = \dfrac{2}{1 + x^2}$

5. $\dfrac{dy}{dx} = 5^x \ln 5 \cdot \log_5 x + \dfrac{5^x}{x \ln 5} = 5^x\left(\ln x + \dfrac{1}{x \ln 5}\right)$

6. $\dfrac{dy}{dx} = 2^{\sin^{-1} x}\,(\ln 2)\,\dfrac{1}{\sqrt{1 - x^2}}$

7. $\dfrac{dy}{dx} = 2 \tan^{-1} 2x + 2x \cdot \dfrac{1}{1 + 4x^2} \cdot 2 - \dfrac{1}{\sqrt{1 + 4x^2}} \cdot \dfrac{8x}{2\sqrt{1 + 4x^2}}$

$= 2 \tan^{-1} 2x$

8. $\dfrac{dy}{dx} = \dfrac{1}{\sqrt{x}}\,(\sec 3\sqrt{x}\,\tan 3\sqrt{x}) \cdot \dfrac{3}{2\sqrt{x}} + \left(\dfrac{1}{2}\,x^{-3/2}\right)\sec 3\sqrt{x}$

$= \dfrac{\sec 3\sqrt{x}}{2x\sqrt{x}}\,(3\sqrt{x}\,\tan 3\sqrt{x} - 1)$

9. $y = (x)^{\sqrt{x}}$ gives $\ln y = \sqrt{x}\,\ln x$

$\dfrac{1}{y}\dfrac{dy}{dx} = \dfrac{1}{2\sqrt{x}}\ln x + \sqrt{x} \cdot \dfrac{1}{x} = \dfrac{1}{2\sqrt{x}}\,(\ln x + 2)$

$\dfrac{dy}{dx} = \dfrac{1}{2\sqrt{x}}\,(\ln x + 2)(x)^{\sqrt{x}}$

10. $e^y \dfrac{dy}{dx} + e^x = e^{x+y}\dfrac{d}{dx}(x + y) = e^{x+y}\left(1 + \dfrac{dy}{dx}\right)$

Hence, $\left(e^y - e^{x+y}\right)\dfrac{dy}{dx} = e^{x+y} - e^x$, or from the original

expression, $-e^x \dfrac{dy}{dx} = e^y$. Thus, $\dfrac{dy}{dx} = -e^{y-x}$.

11. $u = 3x + 5, \quad du = 3\,dx$

$\displaystyle\int \sec^2(3x + 5)\,dx = \dfrac{1}{3}\int \sec^2 u\,du = \dfrac{1}{3}\tan(3x + 5) + C$

12. $\int \dfrac{dx}{25 + 16x^2} = \dfrac{1}{25} \int \dfrac{dx}{1 + \left(\frac{4}{5}x\right)^2} = \dfrac{1}{25} \cdot \dfrac{5}{4} \tan^{-1} \dfrac{4}{5}x + C$

   $\int_{-5/4}^{5/4} \dfrac{dx}{25 + 16x^2} = \dfrac{1}{20} \left(\tan^{-1} 1 - \tan^{-1} (-1)\right) = \dfrac{\pi}{40}$

13. $\int (1 - e^{-x})^2 \, dx = \int (1 - 2e^{-x} + e^{-2x}) dx = x + 2e^{-x} - \dfrac{1}{2}e^{-2x} + C$

14. $\int \dfrac{1 - 2 \cos x}{\sin^2 x} \, dx = \int \left(\csc^2 x - \dfrac{2 \cos x}{\sin^2 x}\right) dx$

   $= - \cot x - \int \dfrac{2 \, du}{u^2} \qquad (u = \sin x)$

   $= - \cot x + \dfrac{2}{\sin x} + C = - \cot x + 2 \csc x + C$

15. $u = \ln 2x, \quad du = \dfrac{1}{2x} \cdot 2 \, dx = \dfrac{1}{x} \, dx$

   $\int \dfrac{(\ln 2x)^3}{x} \, dx = \int u^3 \, du = \dfrac{1}{4} (\ln 2x)^4 + C$

16. $u = e^{-x}, \quad du = -e^{-x} \, dx$

   $\int \dfrac{-du}{\sqrt{1 - u^2}} = \cos^{-1} \left(e^{-x}\right) + C$

17. $u = \ln x, \quad du = \dfrac{1}{x} \, dx$

   $\int \dfrac{2^{\ln x}}{3x} \, dx = \dfrac{1}{3} \int 2^u \, du = \dfrac{1}{3 \ln 2} 2^{\ln x} + C$

   $\int_1^e \dfrac{2^{\ln x}}{3x} \, dx = \dfrac{1}{3 \ln 2} \left(2^{\ln e} - 1\right) = \dfrac{1}{3 \ln 2}$

18. $\lim_{x \to 0} \csc x \sin^{-1} x = \lim_{x \to 0} \dfrac{\sin^{-1} x}{\sin x} \qquad$ (type 0/0)

   $= \lim_{x \to 0} \dfrac{1/\sqrt{1 - x^2}}{\cos x} = \dfrac{1}{1} = 1$

19.   Let $y = x^{1/(1-x)}$ so $\ln y = \dfrac{\ln x}{1 - x}$

   $\lim_{x \to 1} \ln y = \lim_{x \to 1} \dfrac{\ln x}{1 - x}$ (type 0/0) $= \lim_{x \to 1} \dfrac{1/x}{-1} = -1$

   By the continuity of the natural logarithm,
   $\lim_{x \to 1} x^{1/(1-x)} = e^{-1}$.

20. $\lim_{x \to 0^+} \dfrac{e^x - \cos x}{x \sin x}$ (type 0/0) $= \lim_{x \to 0^+} \dfrac{e^x + \sin x}{x \cos x + \sin x} = +\infty$.

21. $\lim_{x \to 0} \dfrac{4^x - 2^x}{x}$ (type 0/0) $= \lim_{x \to 0} \dfrac{4^x \ln 4 - 2^x \ln 2}{1}$

   $= \ln 4 - \ln 2 = \ln 2$.

22.   (a) $-\dfrac{\pi}{3}$ (b) $\dfrac{\pi}{6}$ (c) 0 (d) $\dfrac{\pi}{3}$

23. $\dfrac{\log_3 243}{\log_2 \sqrt[4]{64} + \log_8 8^{-10}} = \dfrac{5}{\frac{1}{4}(6) - 10(1)} = -\dfrac{10}{17}$

24. (a) $\log_a 10 = \log_a 2 + \log_a 5 = 2.09$

    (b) $\log_a 49 = 2 \log_a 7 = 3.54$

    (c) $\log_a \frac{5}{7} = \log_a 5 - \log_a 7 = -0.31$

    (d) $\log_a 1.4 = \log_a 14 - \log_a 10$
    $\qquad\qquad = \log_a 2 + \log_a 7 - \log_a 5 - \log_a 2 = 0.31$

25. (a) $3^{-8x+6} = 3^{-3x-24}$ or $-8x + 6 = -3x - 24$ or $x = 6$.

    (b) $5^{-2} = 5x - 1$ or $x = 26/125$.

    (c) $x = (x + 1) \ln 10$ or $x = \dfrac{\ln 10}{1 - \ln 10} \approx -1.768$.

26. $f(x) = \dfrac{2}{1 + 3e^{-2x}}$, $\qquad f'(x) = \dfrac{12e^{-2x}}{\left(1 + 3e^{-2x}\right)^2}$,

    $f''(x) = \dfrac{24e^{-2x}\left(-1 + 3e^{-2x}\right)}{\left(1 + 3e^{-2x}\right)^3}$

    $f' > 0$ for every x;

    $f''(x) = 0$ when $x = \frac{1}{2} \ln 3 \approx 0.55$

    $\lim\limits_{x \to +\infty} f(x) = 2$ and $\lim\limits_{x \to -\infty} f(x) = 0$

    The graph is sketched at the right.

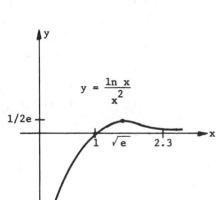

27. $f(x) = \dfrac{\ln x}{x^2}$,

    $f'(x) = \dfrac{1 - 2 \ln x}{x^3}$ so

    $f'(x) = 0$ implies $\ln x = \frac{1}{2}$ or $x = \sqrt{e}$

    $f''(x) = \dfrac{-5 + 6 \ln x}{x^4}$ so

    $f''(x) = 0$ implies $x = e^{5/6} \approx 2.3$

    $\lim\limits_{x \to \infty} f(x) = \lim\limits_{x \to \infty} \dfrac{1/x}{2x} = 0$ and

    $\lim\limits_{x \to 0^+} f(x) = -\infty$. The graph is sketched
    at the right.

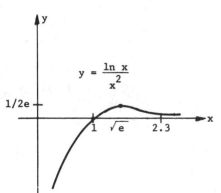

28.   Let x denote the number of bacteria present at any time t. Then, $\frac{dx}{dt} = kx$ or $x = Ce^{kt}$, for some constant C. If x_0 is the initial number of bacteria at t = 0, then $C = x_0$, so $x = x_0\,e^{kt}$. When t = 1, $x = 2x_0$ so that $2 = e^k$ or k = ln 2. Therefore,

$$x = x_0\;e^{t\,\ln 2}.$$

Finally, when t = 3.5, $x = x_0\,e^{3.5\,\ln 2} \approx 11.31\,x_0$. Thus, there are 11.31 times the initial number of bacteria present at the end of 3.5 hours.

CHAPTER 7 METHODS OF INTEGRATION

7-1 BASIC INTEGRATION FORMULAS.

$\boxed{\text{OBJECTIVE}}$: Evaluate indefinite integrals by reducing the integrands to standard forms through simple substitution.

1. $\int x(a + x)^{1/3} \, dx$, where a is constant.

   Let $u = a + x$. Then $du =$ _____ and $x =$ _____. Thus,

   $$\int x(a + x)^{1/3} \, dx = \int \underline{\hspace{1.5cm}} \, du = \int u^{4/3} \, du - \int \underline{\hspace{1.5cm}} \, du$$

   $$= \underline{\hspace{2cm}} - \frac{3}{4} au^{4/3} + C$$

   $$= \underline{\hspace{3cm}} + C.$$

2. $\int \dfrac{dx}{x \sqrt{x - 5}}$

   Let $u = \sqrt{x - 5}$. Then $du =$ _____ and $x =$ _____.

   $$\int \frac{dx}{x \sqrt{x - 5}} = \int \underline{\hspace{1.5cm}} \, du = \underline{\hspace{1cm}} \int \frac{du}{\left(\frac{u}{\sqrt{5}}\right)^2 + 1}$$

   $$= \underline{\hspace{1.5cm}} \tan^{-1} \frac{u}{\sqrt{5}} + C = \underline{\hspace{2cm}} + C.$$

3. $\int \dfrac{\cos x \, dx}{\sqrt{1 - \sin x}}$

   Let $u = \sin x$, then $du =$ _____ so that

   $$\int \frac{\cos x \, dx}{\sqrt{1 - \sin x}} = \int \underline{\hspace{1cm}} \, du = \int (1 - u) \, \underline{\hspace{1cm}} \, du$$

   $$= \underline{\hspace{1.5cm}} + C = \underline{\hspace{2cm}}.$$

4. $\int \dfrac{3^{\tan x} \, dx}{\cos^2 x}$

   Let $u = \tan x$. Then $du =$ _____ and the integral is

   $$\int \frac{3^{\tan x} \, dx}{\cos^2 x} = \int \sec^2 x \; 3^{\tan x} \, dx = \int \underline{\hspace{1cm}} \, du$$

   $$= \underline{\hspace{1.5cm}} + C = \underline{\hspace{2cm}}.$$

---

1. dx, $u - a$, $(u - a)u^{1/3}$, $au^{1/3}$, $\frac{3}{7} u^{7/3}$, $3(a + x)^{4/3} \left[\frac{1}{7}(a + x) - \frac{a}{4}\right]$

2. $\dfrac{dx}{2\sqrt{x - 5}}$, $u^2 + 5$, $\dfrac{2}{u^2 + 5}$, $\dfrac{2}{5}$, $\dfrac{2\sqrt{5}}{5}$, $\dfrac{2}{\sqrt{5}} \tan^{-1} \sqrt{\dfrac{x}{5} - 1}$

3. $\cos x \, dx$, $\dfrac{1}{\sqrt{1 - u}}$, $-\dfrac{1}{2}$, $-2(1 - u)^{1/2}$, $-2\sqrt{1 - \sin x} + C$

4. $\sec^2 x \, dx$, 3^u, $\dfrac{1}{\ln 3} 3^u$, $\dfrac{1}{\ln 3} 3^{\tan x} + C$

5. $\displaystyle\int \frac{dx}{x + x \ln^2 x}$

Let $u = \ln x$. Then $du =$ _____ so that

$$\int \frac{dx}{x + x \ln^2 x} = \int \frac{dx}{x \left(\underline{\hspace{1.5cm}} \right)} = \int \frac{du}{\underline{\hspace{1cm}}}$$

$$= \underline{\hspace{2.5cm}} + C = \underline{\hspace{3cm}}.$$

7-2   INTEGRATION BY PARTS.

OBJECTIVE :   Evaluate integrals by the method of integration by parts whenever feasible.

6. The method of integration by parts is based on the product rule for differentiation. In terms of integrals,

$$\int u \ dv = \underline{\hspace{3cm}}.$$

7. To employ the method of integration by parts successfully we must be able to integrate the part _____ immediately in order to obtain _____. Also it is hoped that the integral $\displaystyle\int v \ du$ is simpler than the original integral $\displaystyle\int u \ dv$.

8. $\displaystyle\int \frac{xe^x \ dx}{(1 + x)^2}$

Let $u = xe^x$, $du =$ _____, $dv = \dfrac{dx}{\left(1 + x\right)^2}$, $v =$ _____.
Then,

$$\int \frac{xe^x \ dx}{(1 + x)^2} = \frac{-xe^x}{1 + x} + \int \underline{\hspace{1.5cm}} \ dx + C$$

$$\int \frac{xe^x \ dx}{(1 + x)^2} = e^x \left(\frac{-x}{1 + x} + \underline{\hspace{1cm}} \right) + C = \underline{\hspace{2.5cm}}.$$

9. $\displaystyle\int \tan^{-1} \sqrt{x} \ dx$

Let $u = \tan^{-1} \sqrt{x}$, $du =$ _____, $dv =$ _____, $v =$ _____.
Then,

$$\int \tan^{-1} \sqrt{x} \ dx = \underline{\hspace{3cm}} - \int \frac{\sqrt{x} \ dx}{2(1 + x)} + C.$$

Let $w = \sqrt{x}$ and $dw =$ _____, and the latter integral becomes

---

5. $\frac{1}{x} dx$, $1 + \ln^2 x$, $1 + u^2$, $\tan^{-1} u$, $\tan^{-1} (\ln x) + C$

6. $uv - \displaystyle\int v \ du + C$ 7. dv, v

8. $e^x(1 + x) dx$, $- \dfrac{1}{1 + x}$, e^x, 1, $\dfrac{e^x}{1 + x} + C$

$$\int \frac{\sqrt{x}\ dx}{2(1+x)} = \int \frac{w^2\ dw}{\underline{\hspace{1cm}}} = \int \left(1 - \underline{\hspace{2cm}}\right) dw$$

$$= w - \underline{\hspace{2cm}}$$ (we may omit the constant of integration here since

$$= \underline{\hspace{3cm}}$$ we already have C)

Putting this together with our previous result, we find

$$\int \tan^{-1} \sqrt{x}\ dx = \underline{\hspace{4cm}}.$$

10. $\int_1^e (\ln x)^2\ dx$

Let $u = (\ln x)^2$, $du = \underline{\hspace{2cm}}$, $dv = \underline{\hspace{2cm}}$,

$v = \underline{\hspace{2cm}}$.   Then,

$$\int_1^e (\ln x)^2\ dx = x(\ln x)^2]_1^e - 2\int_1^e \underline{\hspace{3cm}}$$

$$= (e - \underline{\hspace{1.5cm}}) - 2[\underline{\hspace{2cm}}]_1^e$$
$$\uparrow$$

Example 3, page 452, in the text

$$= e - 2(\underline{\hspace{2cm}}) = \underline{\hspace{2cm}}.$$

11. $\int x^2 \sin x\ dx$

Let $u = x^2$, $du = \underline{\hspace{1.5cm}}$, $dv = \underline{\hspace{1.5cm}}$, $v = \underline{\hspace{1.5cm}}$.   Then,

$$\int x^2 \sin x\ dx = \underline{\hspace{2cm}} + 2\int x \cos x\ dx.$$

To determine the latter integral we integrate by parts again:

$U = x$, $dU = \underline{\hspace{1.5cm}}$, $dV = \underline{\hspace{1.5cm}}$, $V = \underline{\hspace{1.5cm}}$.   Then,

$$\int x \cos x\ dx = x \sin x - \int \underline{\hspace{2cm}} = \underline{\hspace{3cm}} + C$$

Therefore, putting these results together, we find

$$\int x^2 \sin x\ dx = \underline{\hspace{4cm}}.$$

12. $\int \frac{\tan^{-1} x\ dx}{x^2}$

Let $u = \tan^{-1} x$, $du = \underline{\hspace{1.5cm}}$, $dv = \underline{\hspace{1cm}}$, $v = \underline{\hspace{2cm}}$.
Then,

---

9. $\dfrac{dx}{2\sqrt{x}(1+x)}$, dx, x, $x \tan^{-1} \sqrt{x}$, $\dfrac{dx}{2\sqrt{x}}$, $1 + w^2$, $\dfrac{1}{1+w^2}$, $\tan^{-1} w$, $\sqrt{x} - \tan^{-1} \sqrt{x}$,

$(x+1) \tan^{-1} \sqrt{x} - \sqrt{x} + C$

10. $\frac{2}{x} \ln x\ dx$, dx, x, $\ln x\ dx$, 0, $x \ln x - x$, $(e - e + 1)$, $e - 2$

11. $2x\ dx$, $\sin x\ dx$, $- \cos x$, $-x^2 \cos x$, dx, $\cos x\ dx$, $\sin x$, $\sin x\ dx$, $x \sin x + \cos x$,

$(2 - x^2) \cos x + 2x \sin x + C$

$$\int \frac{\tan^{-1} x \ dx}{x^2} = \underline{\hspace{2cm}} + \int \frac{dx}{x(1 + x^2)}$$

$$= \underline{\hspace{2cm}} + \int \frac{dx}{x} + \int \frac{\underline{\hspace{1.5cm}}}{1 + x^2}$$

$$= \underline{\hspace{4cm}} \ .$$

7-3   PRODUCTS AND POWERS OF TRIGONOMETRIC FUNCTIONS
(OTHER THAN EVEN POWERS OF SINES AND COSINES).

OBJECTIVE A :   Evaluate integrals of products and powers of
trigonometric functions.

13. $\int \sin^5 x \ dx = \int \sin^4 x \cdot \underline{\hspace{2cm}} = \int (1 - \cos^2 x) \underline{\hspace{0.8cm}} \sin x \ dx$

Let $u = \underline{\hspace{1.5cm}}$ so that $du = - \sin x \ dx$.   Then,

$$\int (1 - \cos^2 x)^2 \sin x \ dx = \int \underline{\hspace{2cm}} \ du$$

$$= - \int (1 - 2u^2 + \underline{\hspace{1cm}}) \ du = \underline{\hspace{2.5cm}} - \tfrac{1}{5} u^5 + C$$

$$= \underline{\hspace{4cm}} \ .$$

14. $\int \cos^3 x \sin^2 x \ dx = \int \cos^2 x \sin^2 x \underline{\hspace{2cm}}$

$$= \int (1 - \sin^2 x) \sin^2 x \underline{\hspace{1.5cm}} = \int (\sin^2 x - \underline{\hspace{1.5cm}}) \cos x \ dx$$

$$= \int \sin^2 x \cos x \ dx - \int \underline{\hspace{3cm}} \ dx$$

$$= \int u^2 \ du - \int u^4 \ du, \quad \text{where} \quad u = \underline{\hspace{1.5cm}} \quad \text{and} \quad du = \underline{\hspace{1.5cm}},$$

$$= \underline{\hspace{2.5cm}} + C = \tfrac{1}{3} \sin^3 x + \underline{\hspace{2cm}} \ .$$

15. $\int \tan^3 x \ dx = \int \tan^2 x \cdot \underline{\hspace{2.5cm}} = \int (\sec^2 x - 1) \underline{\hspace{2.5cm}}$

$$= \int \underline{\hspace{2.5cm}} \ dx - \int \tan x \ dx$$

$$= \underline{\hspace{4cm}} \ .$$

16. $\int \sec^5 2x \tan 2x \ dx$

Let $u = \sec 2x$.   Then $du = \underline{\hspace{2.5cm}}$ and the integral
becomes

---

12. $\frac{1}{1 + x^2} \ dx$, $\frac{dx}{x^2}$, $- \frac{1}{x}$, $- \frac{\tan^{-1} x}{x}$, $- \frac{\tan^{-1} x}{x}$, $- x \ dx$, $- \frac{\tan^{-1} x}{x} + \ln |x| - \tfrac{1}{2} \ln (1 + x^2) + C$

13. $\sin x \ dx$, 2, $\cos x$, $-\left(1 - u^2\right)^2$, u^4, $- u + \tfrac{2}{3} u^3$, $- \cos x + \tfrac{2}{3} \cos^3 x - \tfrac{1}{5} \cos^5 x + C$

14. $\cos x \ dx$, $\cos x \ dx$, $\sin^4 x$, $\sin^4 x \cos x$, $\sin x$, $\cos x \ dx$, $\tfrac{1}{3} u^3 - \tfrac{1}{5} u^5$, $- \tfrac{1}{5} \sin^5 x + C$

15. $\tan x \ dx$, $\tan x \ dx$, $\tan x \sec^2 x$, $\tfrac{1}{2} \tan^2 x + \ln |\cos x| + C$

$$\int \sec^5 2x \tan 2x \, dx = \int \sec^4 2x \cdot \underline{\hspace{3cm}}$$

$$= \int \underline{\hspace{2cm}} \cdot \tfrac{1}{2} \, du = \underline{\hspace{2cm}} + C$$

$$= \underline{\hspace{4cm}}.$$

17. $\int \dfrac{\sin^5 x \, dx}{\sqrt{\cos x}} = \int \dfrac{\sin^4 x}{\sqrt{\cos x}} \cdot \underline{\hspace{3cm}}$

Let $u = \cos x$. Then $du = \underline{\hspace{3cm}}$. Thus,

$$\int \frac{\sin^5 x \, dx}{\sqrt{\cos x}} = \int \frac{(1 - \cos^2 x) \underline{\hspace{0.5cm}}}{\sqrt{\cos x}} \sin x \, dx$$

$$= - \int \frac{(1 - u^2) \underline{\hspace{0.5cm}}}{u^{1/2}} \, du = - \int (u^{-1/2} - 2u^{3/2} + \underline{\hspace{2cm}}) \, du$$

$$= \underline{\hspace{4cm}} + C$$

$$= \underline{\hspace{4cm}}.$$

18. $\int \dfrac{\sec x \, dx}{\tan^2 x} = \int \dfrac{(\underline{\hspace{1.5cm}}) \sec x}{\sin^2 x} \, dx = \int \dfrac{\underline{\hspace{1cm}}}{\sin^2 x} \, dx.$

Let $u = \sin x$, so that $du = \underline{\hspace{2cm}}$ and the integral becomes

$$\int \frac{\sec x \, dx}{\tan^2 x} = \int \frac{\cos x \, dx}{\sin^2 x} = \int \underline{\hspace{2cm}} \, du$$

$$= \underline{\hspace{2cm}} + C = \underline{\hspace{3cm}}.$$

OBJECTIVE B: Find integrals of the functions $\sin mx \sin nx$, $\cos mx \cos nx$, and $\sin mx \cos nx$, where m and n are positive integers.

19. $\int \sin^2 x \cos 3x \, dx$

$$\int \sin^2 x \cos 3x \, dx = \tfrac{1}{2} \int (1 - \cos 2x) \underline{\hspace{2.5cm}} \, dx$$

$$= \tfrac{1}{2} \int \cos 3x \, dx - \tfrac{1}{2} \int \underline{\hspace{2.5cm}} \, dx$$

$$= \tfrac{1}{2} \int \cos 3x \, dx - \tfrac{1}{4} \int [\underline{\hspace{2.5cm}}] \, dx$$

$$= \underline{\hspace{3.5cm}} - \tfrac{1}{20} \sin 5x + C.$$

---

16. $2 \sec 2x \tan 2x \, dx$, $\sec 2x \tan 2x \, dx$, u^4, $\frac{1}{10} u^5$, $\frac{1}{10} \sec^5 2x + C$

17. $\sin x \, dx$, $- \sin x \, dx$, 2, 2, $u^{7/2}$, $-2u^{1/2} + \frac{4}{5} u^{5/2} - \frac{2}{9} u^{9/2}$,

   $-2 \cos^{1/2} x + \frac{4}{5} \cos^{5/2} x - \frac{2}{9} \cos^{9/2} x + C$

18. $\cos^2 x$, $\cos x$, $\cos x \, dx$, u^{-2}, $-\frac{1}{u}$, $- \csc x + C$

19. $\cos 3x$, $\cos 2x \cos 3x$, $\cos x + \cos 5x$, $\frac{1}{6} \sin 3x - \frac{1}{4} \sin x$

7-4   EVEN POWERS OF SINES AND COSINES.

OBJECTIVE :   Evaluate integrals involving even powers of sin x or
cos x.

20. $\int \sin^2 (1 - 3x) \, dx$

Since $\sin^2 \theta = \frac{1}{2} (1 - \cos 2\theta)$,   we have

$\int \sin^2 (1 - 3x) \, dx = \frac{1}{2} \int ($_____$) \, dx$

$= \frac{1}{2} \int dx - \frac{1}{2} \int$ _____

$= \frac{1}{2} x - \frac{1}{2} \int \cos u \cdot$ _____, $\quad u = 2 - 6x$

$= \frac{1}{2} x +$ _____ $+ C$

$=$ _____.

21. $\int \dfrac{dx}{\cos^4 x}$

$\int \dfrac{dx}{\cos^4 x} = \int \sec^4 x \, dx = \int \sec^2 x \cdot ($_____$) \, dx$

$= \int (1 + \tan^2 x)$ _____

Let $u = \tan x$. Then $du =$ _____ so that

$\int \dfrac{dx}{\cos^4 x} = \int$ _____ $du =$ _____ $+ C$

$=$ _____.

22. $\int \dfrac{dx}{\sin^2 x + \cos^2 x} = \int \csc^2 x$ _____

$= \int \csc^2 x (\tan$—$ x + 1) \, dx$

$= \int ($_____ $+ \csc^2 x) \, dx$

$= \tan x + ($_____$) + C.$

---

20. $1 - \cos 2(1 - 3x)$, $\cos (2 - 6x) \, dx$, $-\frac{1}{6} \, du$, $\frac{1}{12} \sin u$, $\frac{1}{2} x + \frac{1}{12} \sin (2 - 6x) + C$

21. $\sec^2 x$, $\sec^2 x \, dx$, $\sec^2 x \, dx$, $1 + u^2$, $u + \frac{1}{3} u^3$, $\tan x + \frac{1}{3} \tan^3 x + C$

22. $\sec^2 x \, dx$, 2, $\sec^2 x$, $- \cot x$

7-5   TRIGONOMETRIC SUBSTITUTIONS THAT REPLACE $a^2 - u^2$, $a^2 + u^2$,   AND $u^2 - a^2$ BY SINGLE SQUARED TERMS.

| OBJECTIVE A |:   Find indefinite integrals of integrands involving the radicals $\sqrt{a^2 - u^2}$, $\sqrt{a^2 + u^2}$, $\sqrt{u^2 - a^2}$,   and the binomials $a^2 + u^2$, $a^2 - u^2$.

23. $\displaystyle\int \frac{x^3 \, dx}{\sqrt{4 - x^2}}$

Let $x = 2 \sin u$ so that $dx = $ _____,   and the integral becomes

$$\int \frac{x^3 \, dx}{\sqrt{4 - x^2}} = \int \frac{\underline{} \cdot 2 \cos u \, du}{\sqrt{4 - 4 \sin^2 u}} = \int \frac{du}{\sqrt{4 \cos^2 u}}$$

$$= \pm 8 \int \sin^2 u \, du \quad (\pm \text{ depends on sign of } \underline{})$$

$$= 8 \int \sin^3 u \, du, \quad \text{using only the principal value of}$$

$$u = \sin^{-1} \frac{x}{2}$$

$$= 8 \int (1 - \cos^2 u) \, \underline{}$$

$$= 8 \left(- \cos u + \underline{}\right) + C$$

For the substitution $x = 2 \sin u$, finish labeling the diagram at the right. Then, from the diagram,

cos u = _____,   and substitution gives

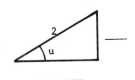

$$\int \frac{x^3 \, dx}{\sqrt{4 - x^2}} = -4\sqrt{4 - x^2} + \underline{} + C.$$

24. $\displaystyle\int \frac{dx}{\left(x^2 - 9\right)^{3/2}}$

Let $x = 3 \sec u$, $dx = $ _____   and the integral becomes

$$\int \frac{dx}{\left(x^2 - 9\right)^{3/2}} = \int \frac{\underline{}}{\left(9 \sec^2 u - 9\right)^{3/2}} = \int \frac{3 \sec u \tan u \, du}{27 \, \underline{}}$$

$$= \frac{1}{9} \int \frac{\sec u \, du}{\underline{}} = \frac{1}{9} \int \frac{\cos u \, du}{\underline{}} = \underline{} + C.$$

For the substitution $x = 3 \sec u$, finish labeling the diagram at the right. Then, from the diagram,

sin u = _____,

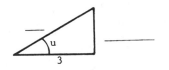

---

23. $2 \cos u \, du$, $8 \sin^3 u$, $8 \sin^3 u \cdot 2 \cos u$,   cos u,   sin u du, $\frac{1}{3} \cos^3 u$,

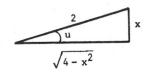

$\frac{1}{2} \sqrt{4 - x^2}$, $\frac{1}{3} \left(4 - x^2\right)^{3/2}$

and substitution gives

$$\int \frac{dx}{\left(x^2 - 9\right)^{3/2}} = \underline{\hspace{4cm}}.$$

25. $\displaystyle\int \frac{dx}{x^2 \sqrt{x^2 + 16}}$

Let   x = 4 tan u,   dx = \underline{\hspace{3cm}}   and the integral becomes

$$\int \frac{dx}{x^2 \sqrt{x^2 + 16}} = \int \frac{4 \sec^2 u \; du}{\underline{\hspace{2cm}}} = \frac{1}{16} \int \frac{\sec u \; du}{\underline{\hspace{2cm}}}$$

$$= - \frac{1}{16} \csc u + C \quad \text{from Problem 18.}$$

For the substitution x = 4 tan u,
finish labeling the diagram at the
right. Then, from the diagram,
csc u = \underline{\hspace{4cm}},
and substitution gives

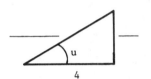

$$\int \frac{dx}{x^2 \sqrt{x^2 + 16}} = \underline{\hspace{4cm}}.$$

$\boxed{\text{OBJECTIVE B}}$:   Calculate definite integrals involving $\sqrt{a^2 - u^2}$,

$\sqrt{a^2 + u^2}$, $\sqrt{u^2 - a^2}$, $a^2 + u^2$,   and $a^2 - u^2$.

26. $\displaystyle\int_{4}^{4\sqrt{3}} \frac{dx}{x^2 \sqrt{x^2 + 16}}$

As in the previous Problem 25, let x = 4 tan u. When x = 4,
tan u = 1 or u = \underline{\hspace{1.5cm}}; when x= $4\sqrt{3}$, tan u = $\sqrt{3}$ or
u = \underline{\hspace{1.5cm}}. Thus, upon substitution,

$$\int_{4}^{4\sqrt{3}} \frac{dx}{x^2 \sqrt{x^2 + 16}} = \frac{1}{16} \int_{\underline{\hspace{0.7cm}}}^{\overline{\hspace{0.7cm}}} \frac{\sec u \; du}{\tan^2 u} \quad \text{(by Problem 25)}$$

$$= - \frac{1}{16} \csc u]\underline{\underline{\hspace{1cm}}}$$

$$= - \frac{1}{16} (\underline{\hspace{3cm}}) \approx 0.016.$$

---

24. 3 sec u tan u du, 3 sec u tan u du, $\tan^3 u$, $\tan^2 u$, $\sin^2 u$, $- \dfrac{1}{9 \sin u}$,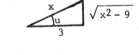

$\dfrac{\sqrt{x^2 - 9}}{x}$, $\dfrac{-x}{9\sqrt{x^2 - 9}} + C$

25. $4 \sec^2 u \; du$, $16 \tan^2 u \cdot 4 \sec u$, $\tan^2 u$,

$\dfrac{\sqrt{x^2 + 16}}{x}$, $-\dfrac{\sqrt{x^2 + 16}}{16x} + C$

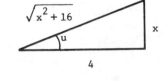

26. $\dfrac{\pi}{4}$, $\dfrac{\pi}{3}$, $\displaystyle\int_{\pi/4}^{\pi/3}$, $]_{\pi/4}^{\pi/3}$, $\dfrac{2\sqrt{3}}{3} - \sqrt{2}$

7-6 INTEGRALS INVOLVING $ax^2 + bx + c$.

[OBJECTIVE]: Find indefinite integrals involving quadratics
$ax^2 + bx + c$.

27. $\int \dfrac{(x + 1)\ dx}{x^2 + x + 5}$

The quadratic part may be transformed algebrically as follows:

$x^2 + x + 5 = (x^2 + x + \underline{\quad}) + 5 - \frac{1}{4} = (\underline{\qquad})^2 + \frac{19}{4}$.

Now, set $u = x + \frac{1}{2}$ and $du = \underline{\qquad}$ so the integral becomes

$$\int \dfrac{(x + 1)dx}{x^2 + x + 5} = \int \dfrac{(x + 1)dx}{\left(x + \frac{1}{2}\right)^2 + \frac{19}{4}} = \int \dfrac{(\underline{\qquad})\ du}{u^2 + \frac{19}{4}}$$

$$= \int \dfrac{}{u^2 + \frac{19}{4}} + \frac{1}{2} \int \dfrac{du}{u^2 + \frac{19}{4}}$$

$$= \int \dfrac{}{z} + \frac{1}{2} \int \dfrac{du}{u^2 + \frac{19}{4}} \quad \left(z = u^2 + \frac{19}{4}\right)$$

$$= \underline{\qquad\qquad} + \frac{1}{2} \sqrt{\frac{4}{19}}\ \tan^{-1} \dfrac{u}{\sqrt{19/4}} + C$$

$$= \frac{1}{2}\ \ln(\underline{\qquad}) + \frac{1}{\sqrt{19}}\ \tan^{-1}\dfrac{2u}{\sqrt{19}} + C$$

$$= \underline{\qquad\qquad\qquad\qquad}.$$

7-7 THE INTEGRATION OF RATIONAL FUNCTIONS—PARTIAL FRACTIONS.

[OBJECTIVE]: Find indefinite integrals of rational functions by the
method of partial fractions expansion.

28. The success of separating a rational function $f(x)/g(x)$ into
a sum of partial fractions hinges upon two things:
(1) The degree of $f(x)$ must be $\underline{\qquad\qquad\qquad\qquad\qquad}$.
If this is not the case, one must first perform
$\underline{\qquad\qquad\qquad}$, then work with the $\underline{\qquad\qquad}$ term.
(2) The factors of $\underline{\qquad}$ must be known. In practice it may
be difficult to perform the factorization.

29. To find $A, B,$ and C in the partial fractions expansion
$$\dfrac{2x^2 - x + 1}{(x + 1)(x - 3)(x + 2)} = \dfrac{A}{x + 1} + \dfrac{B}{x - 3} + \dfrac{C}{x + 2}$$

by the Heaviside technique, the value of A can be found by
covering up the factor $\underline{\qquad}$ in the left side and evaluating
the result at $x = \underline{\qquad}$:

---

27. $\frac{1}{4}$, $x + \frac{1}{2}$, dx, $u + \frac{1}{2}$, $u\ du$, $\frac{1}{2}\ dz$, $\frac{1}{2}\ln|z|$, $u^2 + \frac{19}{4}$, $\frac{1}{2}\ln(x^2 + x + 5) + \frac{1}{\sqrt{19}}\tan^{-1}\dfrac{2x + 1}{\sqrt{19}} + C$

28. less than the degree of $g(x)$, long division, remainder, $g(x)$

$$A = \frac{2(-1)^2 - (-1) + 1}{\underline{\hspace{2cm}}} = \frac{4}{\underline{\hspace{2cm}}} = \underline{\hspace{2cm}} .$$

Similarly, $B = \dfrac{2(3)^2 - (3) + 1}{\underline{\hspace{2cm}}} = \dfrac{16}{\underline{\hspace{2cm}}} = \underline{\hspace{2cm}}$, and

$$C = \frac{\overline{\hspace{2cm}}}{(-2 + 1)(-2 - 3)} = \frac{\overline{\hspace{1cm}}}{5} .$$

30. $\displaystyle\int \frac{(2x^2 - x + 1)\,dx}{(x + 1)(x - 3)(x + 2)} = \int \frac{-\,dx}{\underline{\hspace{1cm}}} + \frac{4}{5} \int \frac{dx}{\underline{\hspace{1cm}}} + \frac{11}{5} \int \frac{dx}{\underline{\hspace{1cm}}}$

   $= \underline{\hspace{6cm}} + C.$

31. $\displaystyle\int \frac{4\,dx}{x^3 + 4x}$

   To expand $\dfrac{4}{x^3 + 4x}$ into a sum of partial fractions, first write,

   $\dfrac{4}{x^3 + 4x} = \dfrac{4}{x\left(x^2 + 4\right)} = \dfrac{A}{x} + \dfrac{\overline{\hspace{1cm}}}{x^2 + 4}$.   Then,

   $4 = A(x^2 + 4) + \underline{\hspace{2cm}} = (\underline{\hspace{1cm}})x^2 + Cx + 4A$.   Thus, equating coefficients of like powers of x, $A + B = \underline{\hspace{1.5cm}}$, $C = \underline{\hspace{1cm}}$, and $4A = \underline{\hspace{1.5cm}}$.   Solving these equations gives $A = \underline{\hspace{1.5cm}}$, $B = \underline{\hspace{1.5cm}}$, and $C = \underline{\hspace{1.5cm}}$.   Hence,

   $\displaystyle\int \frac{4\,dx}{x^3 + 4x} = \int \frac{dx}{\underline{\hspace{1cm}}} - \int \frac{x\,dx}{\underline{\hspace{1cm}}}$

   $= \ln |x| - \underline{\hspace{3cm}} + C = \underline{\hspace{5cm}}.$

32. $\displaystyle\int \frac{dx}{\sin x \cos x}$

   Write $\displaystyle\int \frac{dx}{\sin x \cos x} = \int \frac{\sin x\,dx}{\sin^2 x \cos x} = \int \frac{\sin x\,dx}{(1 - \cos^2 x)\cos x}$,
   and let $u = \cos x$, $du = \underline{\hspace{2cm}}$ so that

   $\displaystyle\int \frac{dx}{\sin x \cos x} = \int \frac{-du}{\underline{\hspace{1.5cm}}}$.   Next, let

   $\dfrac{1}{(1 - u^2)u} = \dfrac{1}{(1 - u)(1 + u)u} = \dfrac{A}{u} + \dfrac{B}{1 - u} + \dfrac{C}{1 + u}$.

   By the Heaviside technique, $A = \underline{\hspace{1.5cm}}$, $B = \underline{\hspace{1.5cm}}$, and $C = \underline{\hspace{1.5cm}}$.   Thus,

---

29. $(x + 1)$, -1, $(-1 - 3)(-1 + 2)$, -4, -1, $(3 + 1)(3 + 2)$, 20, $\frac{4}{5}$, $2(-2)^2 - (-2) + 1$, 11

30. $x + 1$, $x - 3$, $x + 2$, $-\ln|x + 1| + \frac{4}{5}\ln|x - 3| + \frac{11}{5}\ln|x + 2|$

31. $Bx + C$, $x(Bx + C)$, $A + B$, 0, 0, 4, 1, -1, 0, x, $x^2 + 4$, $\frac{1}{2}\ln(x^2 + 4)$, $\ln \dfrac{|x|}{\sqrt{x^2 + 4}} + C$

32. $-\sin x\,dx$, $(1 - u^2)u$, 1, $\frac{1}{2}$, $-\frac{1}{2}$, u, $1 - u$, $1 + u$,

   $-\ln|u| + \frac{1}{2}\ln|1 - u| + \frac{1}{2}\ln|1 + u|$, $\left(1 - u^2\right)^{1/2}$, $|\cos x|$, $\ln|\tan x| + C'$

$$\int \frac{-\,du}{(1 - u^2)u} = -\int \frac{du}{\underline{\hspace{1.5cm}}} - \frac{1}{2}\int \frac{du}{\underline{\hspace{1.5cm}}} + \frac{1}{2}\int \frac{du}{\underline{\hspace{1.5cm}}}$$

$$= \underline{\hspace{6cm}} + C'$$

$$= \ln \frac{1}{|u|}\,[\underline{\hspace{3cm}}] + C'$$

$$= \ln \frac{\sqrt{1 - \cos^2 x}}{\underline{\hspace{1.5cm}}} + C' = \underline{\hspace{5cm}}.$$

33. $\displaystyle\int \frac{(x^3 - x + 2)\,dx}{x^2 - x}$

Here the degree of the numerator fails to be less than the degree of the denominator so we must use long division and work with the remainder.

By long division, $\dfrac{x^3 - x + 2}{x^2 - x} = x + 1 + \underline{\hspace{2cm}}$.   Therefore,

$$\int \frac{(x^3 - x + 2)\,dx}{x^2 - x} = \int (x + 1)\,dx + 2\int \frac{dx}{\underline{\hspace{1.5cm}}}.$$

By partial fractions, $\dfrac{1}{x(x - 1)} = \dfrac{A}{x} + \dfrac{B}{x - 1}$,   and the Heaviside

technique gives $A = \underline{\hspace{2cm}}$ and $B = \underline{\hspace{2cm}}$.   Therefore,

$$\int \frac{(x^3 - x + 2)\,dx}{x^2 - x} = \underline{\hspace{3cm}} + 2\int \frac{dx}{\underline{\hspace{1.5cm}}} - 2\int \frac{dx}{\underline{\hspace{1.5cm}}}$$

$$= \frac{x^2}{2} + x + 2\,(\underline{\hspace{4cm}}) + C$$

$$= \underline{\hspace{6cm}}.$$

34. $\displaystyle\int \frac{(2x^2 + x + 2)\,dx}{x(x - 1)^2}$

By partial fractions,

$$\frac{2x^2 + x + 2}{x(x - 1)^2} = \frac{A}{x} + \frac{B}{x - 1} + \underline{\hspace{3cm}}.$$

Clearing fractions and equating numerators gives,

$$2x^2 + x + 2 = A(x - 1)^2 + Bx(x - 1) + \underline{\hspace{2.5cm}}$$

$$= (A + B)x^2 + (\underline{\hspace{2.5cm}})x + A$$

Equating coefficients of like powers of x gives the equations
$A + B = 2$, $\underline{\hspace{3cm}} = 1$, and $\underline{\hspace{2.5cm}} = 2$. Solving
simultaneously, $A = \underline{\hspace{1.5cm}}$, $B = \underline{\hspace{1.5cm}}$, and $C = 5$. Hence,

$$\int \frac{(2x^2 + x + 2)\,dx}{x(x - 1)^2} = 2\int \frac{dx}{\underline{\hspace{1.5cm}}} + 5\int \frac{dx}{\underline{\hspace{1.5cm}}} = \underline{\hspace{4cm}}.$$

---

33. $\dfrac{2}{x^2 - x}$, $x(x - 1)$, -1, 1, $\frac{1}{2}x^2 + x$, $x - 1$, x, $\ln|x - 1| - \ln|x|$, $\dfrac{x^2}{2} + x + 2\ln\left|\dfrac{x - 1}{x}\right| + C$

34. $\dfrac{C}{(x - 1)^2}$, Cx, $-2A - B + C$, $-2A - B + C$, A, 2, 0, x, $(x - 1)^2$, $2\ln|x| - \dfrac{5}{x - 1} + C$

7-8   IMPROPER INTEGRALS.

35.   A definite integral $\int_a^b f(x)\,dx$ is termed _improper_ if,

(a) either limit of integration is _____ or _____, or
(b) $f(x)$ becomes _____ at some value $x = c$ satisfying
_____ .

$\boxed{\text{OBJECTIVE}}$: Determine whether a given improper integral converges or diverges. If convergent, evaluate it.

36. $\displaystyle \int_1^\infty \frac{\ln x}{x^2}\,dx = \lim_{b \to \infty} \int_{\underline{\ \ \ }}^{\underline{\ \ \ }} \frac{\ln x}{x^2}\,dx$

To calculate the latter integral, we integrate by parts: let $u = \ln x$, $du = $ _____, $dv = $ _____, $v = $ _____.

$\displaystyle \int_1^b \frac{\ln x\,dx}{x^2} = -\left.\frac{\ln x}{x}\right]_1^b + \int_1^b \underline{\hspace{2cm}} = \frac{-\ln b}{b} + \underline{\hspace{1.5cm}}$; whence

$\displaystyle \int_1^\infty \frac{\ln x\,dx}{x^2} = \lim_{b \to \infty} \left(\underline{\hspace{2cm}} \right) = \lim_{b \to \infty} \frac{-\ln b}{b} + \underline{\hspace{1.5cm}}$

$\displaystyle \qquad\qquad = \lim_{b \to \infty} \underline{\hspace{1.5cm}} + 1 = \underline{\hspace{1cm}}$.

37. $\displaystyle \int_0^1 \ln x\,dx$

Here, $\ln x \to$ _____ as $x \to 0^+$ making the integral improper. Thus,

$\displaystyle \int_0^1 \ln x\,dx = \lim_{b \to 0^+} \int_{\underline{\ \ \ }}^{\underline{\ \ \ }} \ln x\,dx.$

To calculate the latter integral, we integrate by parts: let $u = \ln x$, $du = $ _____, $dv = $ _____, $v = $ _____.

$\displaystyle \int_b^1 \ln x\,dx = \left.\underline{\hspace{2cm}}\right]_b^1 - \int_b^1 \underline{\hspace{1cm}} dx = \underline{\hspace{2.5cm}}$. Therefore,

$\displaystyle \int_0^1 \ln x\,dx = \lim_{b \to 0^+} (-b \ln b - 1 + b)$

$\displaystyle \qquad\qquad = \lim_{b \to 0^+} \left(\frac{-\ln b}{1/b} \right) - 1$

$\displaystyle \qquad\qquad = \lim_{b \to 0^+} \left(\underline{\hspace{2cm}} \right) - 1 = \underline{\hspace{1.5cm}}$.

38. $\displaystyle \int_0^\infty \frac{x\,dx}{\sqrt{1 + x^3}}$

Observe that $\displaystyle \frac{x}{\sqrt{1 + x^3}} > \frac{1}{x}$ is true whenever $x^2 > \underline{\hspace{2.5cm}}$

---

35. $+\infty$, $-\infty$, infinite, $a \leq c \leq b$

36. $\displaystyle \int_1^b$, $\frac{1}{x}\,dx$, $\frac{1}{x^2}\,dx$, $-\frac{1}{x}$, $\frac{dx}{x^2}$, $-\frac{1}{x}\big]_1^b$, $\frac{-\ln b}{b} - \frac{1}{b} + 1$, 1, $\frac{-1/b}{1}$ (l'Hôpital's rule), 1

37. $-\infty$, $\int_b^1$, $\frac{dx}{x}$, dx, x, $x \ln x$, $-b \ln b - 1 + b$, $\frac{1/b}{1/b^2}$ (l'Hôpital's rule), -1

whenever $x^4 >$ _____ ,   and the last inequality certainly holds if $x \geq 2$. Thus, we consider the improper integral

$\int_2^\infty \frac{dx}{x}$. Now, $\int_2^\infty \frac{dx}{x} = \lim_{b \to \infty} [\ln b - \ln 2] = +\infty$. Thus, since the original integral satisfies the inequalities,

$\int_0^\infty \frac{x \, dx}{\sqrt{1 + x^3}} \geq \int_2^\infty \frac{x \, dx}{\sqrt{1 + x^3}} \geq \int_2^\infty \frac{dx}{x}$, it _____ to $+\infty$.

39. $\int_0^2 \frac{(x^2 - 3x + 1) \, dx}{x(x - 1)^2}$

In this case the integrand becomes infinite when $x =$ _____ and $x =$ _____,   so the integrand is improper. To simplify the notation momentarily, let $f(x)$ denote the integrand. Then,

$$\int_0^2 f(x) \, dx = \lim_{b \to 0^+} \int_b^{1/2} f(x) \, dx + \lim_{c \to 1^-} \int_{1/2}^c f(x) \, dx$$
$$+ \lim_{h \to 1^+} \int_{\underline{\quad}}^{\underline{\quad}} \underline{\qquad\qquad}$$

To find the indefinite integral $\int \frac{(x^2 - 3x + 1) \, dx}{x(x - 1)^2}$ we expand the integrand by partial fractions:

$$\frac{x^2 - 3x + 1}{x(x - 1)^2} = \frac{A}{x} + \frac{B}{x - 1} + \frac{C}{(x - 1)^2}$$

Therefore,

$$x^2 - 3x + 1 = A(x - 1)^2 + Bx(x - 1) + \underline{\qquad}$$
$$= (A + B)x^2 + (\underline{\qquad})x + A.$$

It follows that $A = 1$, $A + B =$ _____ and _____ $= -3$. Solving simultaneously, $B =$ _____ and $C =$ _____. Hence,

$$\int \frac{(x^2 - 3x + 1) \, dx}{x(x - 1)^2} = \int \frac{dx}{x} + \int \underline{\qquad} = \ln |x| + \underline{\qquad} + C.$$

Therefore,

$$\int_0^2 \frac{(x^2 - 3x + 1) \, dx}{x(x - 1)^2} = \lim_{b \to 0^+} (\underline{\qquad})]_b^{1/2}$$
$$+ \lim_{c \to 1^-} (\underline{\qquad})]_{1/2}^c + \lim_{h \to 1^+} (\underline{\qquad})]_h^2$$

The first limit is _____, the second limit is _____, and the third is _____. Therefore, the improper integral is _____.

---

38. $\sqrt{1 + x^3}$ (since $x > 0$), $1 + x^3$, diverges

39. 0, 1, $\int_h^2 f(x) \, dx$, Cx, $-2A - B + C$, 1, $-2A - B + C$, 0, -1, $\frac{-dx}{(x - 1)^2}$, $\frac{1}{x - 1}$, $\ln x + \frac{1}{x - 1}$,

$\ln x + \frac{1}{x - 1}$, $\ln x + \frac{1}{x - 1}$, $+\infty$, $-\infty$, $-\infty$, divergent

40.   The improper integrals in Problems 36 and 37 could be implicitly evaluated with the aid of indefinite integrals. Such is not the case, however, for the improper integral

$$\int_0^\infty \frac{\sin x \, dx}{1 + x^2} = \lim_{b \to \infty} \int_{\underline{\quad}}^{\underline{\quad}} \frac{\sin x \, dx}{1 + x^2} \leq \lim_{b \to \infty} \int_0^b \frac{dx}{1 + x^2}$$

Evaluating the latter integral,

$$\lim_{b \to \infty} \int_0^b \frac{dx}{1 + x^2} = \lim_{b \to \infty} \underline{\qquad\qquad} \Big]_0^b = \underline{\qquad\qquad}.$$

Therefore, we know that the original integral $\displaystyle\int_0^\infty \frac{\sin x \, dx}{1 + x^2}$ is

_____.

7-9   USING INTEGRAL TABLES.

OBJECTIVE :   Find indefinite integrals with the aid of integral tables.

41. $\displaystyle\int \frac{dx}{x^2 \sqrt{4 + x^2}}$

We use Formula 27 with   a = _____.   Thus,

$$\int \frac{dx}{x^2 \sqrt{4 + x^2}} = \underline{\qquad\qquad} + C.$$

42. $\displaystyle\int x^2 \ln 3x \, dx$

We use Formula 110 with   n = _____   and   a = _____.   Thus,

$$\int x^2 \ln 3x \, dx = \underline{\qquad\qquad\qquad} + C.$$

7-10   REDUCTION FORMULAS.

OBJECTIVE :   Use reduction formulas provided in integral tables to evaluate integrals.

43. $\displaystyle\int (x^2 - 16)^{3/2} \, dx$

We use Formula 38 with   n = _____,   and   a = _____.   Thus,

$$\int (x^2 - 16)^{3/2} \, dx = \frac{x\left(\sqrt{x^2 - 16}\right)^3}{4} - \underline{\qquad\qquad\qquad}.$$ We

next use Formula 29 to evaluate the latter integral and obtain the result

$$\int (x^2 - 16)^{3/2} \, dx = \underline{\qquad\qquad\qquad\qquad} + C.$$

---

40. $\int_0^b$, $\tan^{-1} x$, $\frac{\pi}{2}$, convergent   41. 2, $-\dfrac{\sqrt{4 + x^2}}{4x}$ 42. 2, 3, $\dfrac{x^3}{3} \ln 3x - \dfrac{x^3}{9}$

43. 3, 4, $\dfrac{48}{4} \int \sqrt{x^2 - 16} \, dx$, $\dfrac{x}{4}(x^2 - 16)^{3/2} + 12\left[\dfrac{x}{2}\sqrt{x^2 - 16} + 8 \sin^{-1} \dfrac{x}{4}\right]$

CHAPTER 7   SELF-TEST

Evaluate the integrals in Problems 1-14.

1. $\displaystyle\int \frac{x \; dx}{x^4 + 1}$

2. $\displaystyle\int \frac{(5x + 3) \; dx}{x^3 - 2x^2 - 3x}$

3. $\displaystyle\int \frac{\ln x \; dx}{(x + 1)^2}$

4. $\displaystyle\int \frac{\sqrt{9 - 4x^2} \; dx}{x}$

5. $\displaystyle\int \frac{e^{3x/2} \; dx}{1 + e^{3x/4}}$

6. $\displaystyle\int \csc^3 x \; dx$ (Tables)

7. $\displaystyle\int \sin^3 x \cos^4 x \; dx$

8. $\displaystyle\int \cot^5 x \; dx$

9. $\displaystyle\int \frac{x \; dx}{x^2 + x + 1}$

10. $\displaystyle\int \cos^2 x \sin 2x \; dx$

11. $\displaystyle\int_{-\pi/3}^{\pi/4} x \sec^2 x \; dx$

12. $\displaystyle\int_0^1 \frac{(2x^3 + x + 3) \; dx}{x^2 + 1}$

13. $\displaystyle\int_5^{5\sqrt{3}} \frac{dx}{x\sqrt{25 + x^2}}$

14. $\displaystyle\int_0^a \ln (a^2 + x^2) \; dx, \quad a > 0$

In Problems 15-18, determine the convergence or divergence of the integral. If the integral is convergent, find its value.

15. $\displaystyle\int_0^{\pi/2} \sec x \tan x \; dx$

16. $\displaystyle\int_{-1}^{1} \frac{dx}{x^{2/3}}$

17. $\displaystyle\int_1^{\infty} \frac{\ln x \; dx}{x}$

18. $\displaystyle\int_0^{\infty} \frac{\sin x \; dx}{e^x}$

SOLUTIONS TO CHAPTER 7 SELF-TEST

1. Let $u = x^2$, $du = 2x \; dx$. Then,

$$\int \frac{x \; dx}{x^4 + 1} = \frac{1}{2} \int \frac{du}{u^2 + 1} = \frac{1}{2} \tan^{-1} u + C = \frac{1}{2} \tan^{-1} x^2 + C.$$

2. $\dfrac{5x + 3}{x^3 - 2x^2 - 3x} = \dfrac{5x + 3}{x(x + 1)(x - 3)} = \dfrac{A}{x} + \dfrac{B}{x + 1} + \dfrac{C}{x - 3}$

By the Heaviside technique, $A = -1$, $B = -\frac{1}{2}$, $C = \frac{3}{2}$. Thus,

$$\int \frac{(5x + 3) \; dx}{x^3 - 2x^2 - 3x} = -\int \frac{dx}{x} - \frac{1}{2} \int \frac{dx}{x + 1} + \frac{3}{2} \int \frac{dx}{x - 3}$$

$$= -\ln |x| - \frac{1}{2} \ln |x + 1| + \frac{3}{2} \ln |x - 3| + C$$

$$= \ln \sqrt{\frac{(x - 3)^3}{x^2(x + 1)}} + C$$

3.   Let $u = \ln x$, $du = \frac{dx}{x}$, $dv = \frac{dx}{(x + 1)^2}$, $v = \frac{-1}{x + 1}$; $x > 0$

$$\int \frac{\ln x \, dx}{(x + 1)^2} = -\frac{\ln x}{x + 1} + \int \frac{dx}{x(x + 1)}$$

$$= -\frac{\ln x}{x + 1} + \int \frac{dx}{x} - \int \frac{dx}{x + 1}$$

$$= -\frac{\ln x}{x + 1} + \ln x - \ln (x + 1) + C$$

$$= -\frac{\ln x}{x + 1} + \ln \frac{x}{x + 1} + C, \quad x > 0.$$

4.   Let $x = \frac{3}{2} \sin u$, $dx = \frac{3}{2} \cos u \, du$.   Then,

$$\int \frac{\sqrt{9 - 4x^2}}{x} \, dx = \int \frac{\sqrt{9 - 9 \sin^2 u} \cdot \frac{3}{2} \cos u \, du}{\frac{3}{2} \sin u}$$

$$= 3 \int \frac{\cos^2 u \, du}{\sin u} = 3 \int \frac{(1 - \sin^2 u) \, du}{\sin u}$$

$$= 3 \int (\csc u - \sin u) \, du$$

$$= 3 \ln |\csc u - \cot u| + 3 \cos u + C$$

From the substitution $x = \frac{3}{2} \sin u$,
and the diagram at the right,

$\csc u = \frac{3}{2x}$ and $\cot u = \frac{\sqrt{9 - 4x^2}}{2x}$.   Thus,

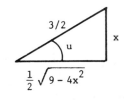

$$\int \frac{\sqrt{9 - 4x^2}}{x} \, dx = 3 \ln \left| \frac{3}{2x} - \frac{\sqrt{9 - 4x^2}}{2x} \right| + \sqrt{9 - 4x^2} + C.$$

5.   Let $u = e^x$, $du = e^x \, dx$.   Then,

$$\int \frac{e^{3x/2} \, dx}{1 + e^{3x/4}} = \int \frac{e^{x/2} \cdot e^x \, dx}{1 + e^{3x/4}} = \int \frac{u^{1/2} \, du}{1 + u^{3/4}}$$

Next, let $z = u^{1/4}$, $dz = \frac{1}{4} u^{-3/4} \, du$.   Hence, $z^2 = u^{1/2}$,
$z^3 = e^{3/4}$,   and $du = 4u^{3/4} \, dz = 4z^3 \, dz$.   Substitution into the
last integral gives

$$\int \frac{e^{3x/2} \, dx}{1 + e^{3x/4}} = \int \frac{z^2(4z^3 \, dz)}{1 + z^3} = 4 \int \frac{z^5 \, dz}{z^3 + 1} = 4 \int \left(z^2 - \frac{z^2}{z^3 + 1} \right) dz$$

$$= 4 \int \left(z^2 - \frac{z^2}{z^3 + 1} \right) dz$$

$$= \frac{4}{3} z^3 - \frac{4}{3} \ln |z^3 + 1| + C$$

$$= \frac{4}{3} u^{3/4} - \frac{4}{3} \ln |u^{3/4} + 1| + C$$

$$= \frac{4}{3} e^{3x/4} - \frac{4}{3} \ln (e^{3x/4} + 1) + C$$

6. $\int \csc^3 x \, dx$

We use the reduction Formula 93 in the integral tables:

$$\int \csc^3 x \, dx = \frac{-\csc x \cot x}{2} + \frac{1}{2} \int \csc x \, dx$$

The last integral is evaluated using Formula 89 to yield

$$\int \csc^3 x \, dx = -\frac{1}{2} \csc x \cot x - \frac{1}{2} \ln |\csc x + \cot x| + C.$$

7. Let $u = \cos x$, $du = -\sin x \, dx$. Then

$$\int \sin^3 x \cos^4 x \, dx = \int \sin^2 x \cos^4 x \sin x \, dx$$

$$= \int (1 - \cos^2 x) \cos^4 x \sin x \, dx$$

$$= \int -(1 - u^2) u^4 \, du$$

$$= -\frac{1}{5} u^5 + \frac{1}{7} u^7 + C = -\frac{1}{5} \cos^5 x + \frac{1}{7} \cos^7 x + C.$$

8. Using the trigonometric identity $\cot^2 x = \csc^2 x - 1$

$$\int \cot^5 x \, dx = \int \cot^3 x \, (\csc^2 x - 1) \, dx$$

$$= \int \cot^3 x \csc^2 x \, dx - \int \cot^3 x \, dx$$

$$= \int \cot^3 x \csc^2 x \, dx - \int \cot x \, (\csc^2 x - 1) \, dx$$

$$= \int \cot^3 x \csc^2 x \, dx - \int \cot x \csc^2 x \, dx + \int \cot x \, dx$$

Let $u = \cot x$, $du = -\csc^2 x \, dx$ and we have

$$\int \cot^5 x \, dx = -\frac{1}{4} \cot^4 x + \frac{1}{2} \cot^2 x + \ln |\sin x| + C.$$

9. Completing the square, $x^2 + x + 1 = \left(x + \frac{1}{2}\right)^2 + \frac{3}{4}.$ Let
    $u = x + \frac{1}{2}$, $du = dx$. Then,

$$\int \frac{x \, dx}{x^2 + x + 1} = \int \frac{\left(u - \frac{1}{2}\right) du}{u^2 + \frac{3}{4}} = \int \frac{u \, du}{u^2 + \frac{3}{4}} - \frac{1}{2} \int \frac{du}{u^2 + \frac{3}{4}}$$

$$= \frac{1}{2} \ln \left|u^2 + \frac{3}{4}\right| - \frac{1}{2} \cdot \frac{2}{\sqrt{3}} \tan^{-1} \frac{2u}{\sqrt{3}} + C$$

$$= \frac{1}{2} \ln |x^2 + x + 1| - \frac{1}{\sqrt{3}} \tan^{-1} \frac{2x + 1}{\sqrt{3}} + C.$$

10. $\int \cos^2 x \sin 2x \, dx = \frac{1}{2} \int (1 + \cos 2x) \sin 2x \, dx$

$$= \frac{1}{2} \int \sin 2x \, dx + \frac{1}{2} \int \sin 2x \cos 2x \, dx$$

$$= \frac{1}{2} \int \sin 2x \, dx + \frac{1}{4} \int \sin 4x \, dx$$

$$= -\frac{1}{4} \cos 2x - \frac{1}{16} \cos 4x + C.$$

11.   Let $u = x$, $du = dx$, $dv = \sec^2 x\, dx$, $v = \tan x$.   Then,

$$\int_{-\pi/3}^{\pi/4} x \sec^2 x\, dx = x \tan x\Big]_{-\pi/3}^{\pi/4} - \int_{-\pi/3}^{\pi/4} \tan x\, dx$$

$$= \frac{\pi}{4} \cdot 1 + \frac{\pi}{3} \cdot \sqrt{3} - \ln |\cos x|\Big]_{-\pi/3}^{\pi/4}$$

$$= \frac{\pi}{4} + \frac{\pi}{\sqrt{3}} - \ln \frac{\sqrt{2}}{2} + \ln \frac{1}{2}$$

$$= \frac{\pi}{4} + \frac{\pi}{\sqrt{3}} - \frac{1}{2} \ln 2 \approx 2.253.$$

12. $\displaystyle\int_0^1 \frac{(2x^3 + x + 3)\, dx}{x^2 + 1} = \int_0^1 \left(2x + \frac{3 - x}{x^2 + 1}\right) dx$

$$= \int_0^1 2x\, dx + 3 \int_0^1 \frac{dx}{x^2 + 1} - \int_0^1 \frac{x\, dx}{x^2 + 1}$$

$$= x^2 + 3 \tan^{-1} x - \frac{1}{2} \ln (x^2 + 1)\Big]_0^1$$

$$= \left(1 + \frac{3\pi}{4} - \frac{1}{2} \ln 2\right) - (0 + 0 - 0) \approx 3.01.$$

13.   Let $x = 5 \tan u$, $dx = 5 \sec^2 u\, du$.   When $x = 5$, $\tan u = 1$ or $u = \pi/4$;   when $x = 5\sqrt{3}$, $\tan u = \sqrt{3}$ or $u = \pi/3$.   Thus,

$$\int_5^{5\sqrt{3}} \frac{dx}{x\sqrt{25 + x^2}} = \int_{\pi/4}^{\pi/3} \frac{5 \sec^2 u\, du}{5 \tan u \sqrt{25 + 25 \tan^2 u}}$$

$$= \frac{1}{5} \int_{\pi/4}^{\pi/3} \frac{\sec u\, du}{\tan u} = \frac{1}{5} \int_{\pi/4}^{\pi/3} \csc u\, du$$

$$= \frac{1}{5} \ln |\csc u - \cot u|\Big]_{\pi/4}^{\pi/3}$$

$$= \frac{1}{5} \left(\ln \left|\frac{2\sqrt{3}}{3} - \frac{\sqrt{3}}{3}\right| - \ln |\sqrt{2} - 1|\right)$$

$$= \frac{1}{5} \ln \left|\frac{\sqrt{3}}{3(\sqrt{2} - 1)}\right| \approx 0.066.$$

14.   Let $u = \ln\left(a^2 + x^2\right)$, $du = \dfrac{2x\, dx}{a^2 + x^2}$, $dv = dx$,   and $v = x$.   Then,

$$\int_0^a \ln\left(a^2 + x^2\right) dx = x \ln\left(a^2 + x^2\right)\Big]_0^a - \int_0^a \frac{2x^2\, dx}{a^2 + x^2}$$

$$= a \ln 2a^2 - 2 \int_0^a \left(1 - \frac{a^2}{a^2 + x^2}\right) dx$$

$$= a \ln 2a^2 - (2x + 2a \tan^{-1} \tfrac{x}{a}\,]_0^a$$

$$= a \ln 2a^2 - 2a + \frac{2\pi a}{4}.$$

15. $\displaystyle\int_0^{\pi/2} \sec x \tan x\, dx = \lim_{b \to \frac{\pi}{2}^-} \int_0^b \sec x \tan x\, dx$

$$= \lim_{b \to \frac{\pi}{2}^-} \sec x\Big]_0^b = \lim_{b \to \frac{\pi}{2}^-} (\sec b - 1) = \infty.$$

Therefore, the improper integral diverges.

16. $\displaystyle\int_{-1}^{1} \frac{dx}{x^{2/3}} = \lim_{a\to 0^-} \int_{-1}^{a} \frac{dx}{x^{2/3}} + \lim_{a\to 0^+} \int_{a}^{1} \frac{dx}{x^{2/3}}$

$\displaystyle = \lim_{a\to 0^-} 3x^{1/3}\big]_{-1}^{a} + \lim_{a\to 0^+} 3x^{1/3}\big]_{a}^{1}$

$\displaystyle = \lim_{a\to 0^-} \left(3a^{1/3} + 3\right) + \lim_{a\to 0^+} \left(3 - 3a^{1/3}\right) = 6.$

17. For $x \geq e$, $\ln x \geq 1$. Hence,

$\displaystyle\int_{1}^{\infty} \frac{\ln x \; dx}{x} \geq \int_{e}^{\infty} \frac{\ln x \; dx}{x} \geq \int_{e}^{\infty} \frac{dx}{x}.$

Now, $\displaystyle\int_{e}^{\infty} \frac{dx}{x} = \lim_{b\to\infty} \ln x\big]_{e}^{b} = \lim_{b\to\infty} \ln b - 1 = \infty.$

Therefore, the integral $\displaystyle\int_{1}^{\infty} \frac{\ln x \; dx}{x}$ diverges.

18. $\displaystyle\int_{0}^{\infty} \frac{\sin x \; dx}{e^{x}} = \lim_{b\to\infty} \int_{0}^{b} e^{-x} \sin x \; dx$

Let $u = e^{-x}$, $du = -e^{-x}\, dx$, $dv = \sin x \; dx$, $v = -\cos x$.

$\displaystyle\int_{0}^{b} e^{-x} \sin x \; dx = -e^{-x} \cos x\big]_{0}^{b} - \int_{0}^{b} e^{-x} \cos x \; dx$

Let $U = e^{-x}$, $dU = -e^{-x}\, dx$, $dV = \cos x \; dx$, $V = \sin x$, and

$\displaystyle\int_{0}^{b} e^{-x} \cos x \; dx = e^{-x} \sin x\big]_{0}^{b} + \int_{0}^{b} e^{-x} \sin x \; dx.$

Putting these results together,

$\displaystyle 2\int_{0}^{b} e^{-x} \sin x \; dx = -e^{-x} \cos x - e^{-x} \sin x\big]_{0}^{b}$

$\displaystyle = -e^{-b} \cos b + 1 - e^{-b} \sin b$

Now, $\displaystyle\lim_{b\to\infty} \frac{\cos b}{e^{b}} = \lim_{b\to\infty} \frac{\sin b}{e^{b}} = 0.$ Thus,

$\displaystyle\int_{0}^{\infty} \frac{\sin x \; dx}{e^{x}} = \lim_{b\to\infty} \int_{0}^{b} e^{-x} \sin x \; dx = \tfrac{1}{2}.$

NOTES.

CHAPTER 8 CONIC SECTIONS AND OTHER PLANE CURVES

INTRODUCTION

1. The "standard" conic sections are the curves in which a plane cuts a _____ .

2. The "degenerate" conic sections are obtained by passing the plane of intersection through the _____ .

8-1 EQUATIONS FROM THE DISTANCE FORMULA.

$\boxed{\text{OBJECTIVE}}$: Use the distance formula to derive an equation for a set of points $P(x,y)$ that satisfy some specified condition.

3. Find an equation for the points $P(x,y)$ whose distances from $F_1(0,-9)$ and $F_2(0,9)$ have the constant sum 30.
   <u>Solution.</u> The distance from P to F_1 is _____, and from P to F_2 is _____ . Thus, the specified conditions yield the equation
   $$\sqrt{x^2 + (y - 9)^2} + \underline{\hspace{4cm}} = 30 .$$
   Transposing the second term on the left, to the right side, and squaring both sides gives
   $$x^2 + (y - 9)^2 = 900 - 60\sqrt{x^2 + (y + 9)^2} + \underline{\hspace{3cm}}$$
   or, simplifying algebraically,
   $$\sqrt{x^2 + (y + 9)^2} = -\tfrac{1}{10} (\underline{\hspace{3cm}}) .$$
   Squaring both sides of this last equation we find,
   $$x^2 + y^2 + 18y + 81 = \tfrac{1}{100} (\underline{\hspace{4cm}}) ,$$
   or simplifying algebraically, $100x^2 + 64y^2 = \underline{\hspace{2cm}}$.
   Alternatively,
   $$\frac{x^2}{144} + \underline{\hspace{2cm}} = 1 .$$

4. Find an equation for the points $P(x,y)$ whose distances from the line $y = -3$ are always 4 units less than their distance from the point $F(0,7)$.
   <u>Solution.</u> The distance from P to the line is _____ , and from P to F is _____ . Thus, the specified

---

1. cone   2. cone's vertex   3. $\sqrt{x^2 + (y + 9)^2}$, $\sqrt{x^2 + (y - 9)^2}$, $\sqrt{x^2 + (y + 9)^2}$,

$x^2 + (y + 9)^2$, $-6y - 150$, $36y^2 + 1800y + 22500$, 14400, $\frac{y^2}{225}$

condition yields the equation $\sqrt{(y + 3)^2}$ = _____,
or $\sqrt{x^2 + (y - 7)^2}$ = _____.   Squaring both sides
$x^2 + y^2 - 14y + 49$ = _____,   or
$x^2 - 20y + 24$ = _____.   If $y + 3 \geq 0$,   or
$y \geq -3$,   this equation gives $x^2 - 20y + 24$ = _____,   or
simplifying algebraically, x^2 = _____.   On the other
hand, if $y + 3 < 0$,   or $y < -3$,   the equation gives
$x^2 - 20y + 24$ = _____   or x^2 = _____.   But for $y < -3$
this last equality would say that x^2 is a negative number.
This is clearly impossible so   y   must be greater than or equal
to   -3,   and a valid equation for the points $P(x,y)$ is
_____.

8-2   CIRCLES.

OBJECTIVE A :   Find an equation for a circle whose center and radius
are known.

5.   An equation of the circle with center $C(h,k)$ and radius   r
is given by _____.

6.   An equation of the circle with center $C(-1,3)$ and   radius
$r = \sqrt{11}$ is _____   or
$$x^2 + y^2 + 2x - 6y + (\underline{\hspace{1cm}}) = 0.$$

OBJECTIVE B :   Given an equation representing a circle, find the
coordinates of its center and the radius.

7.   Consider the circle $2x^2 + 2y^2 - 8x + 5y + 8 = 0$.   Transposing
the constant term and dividing by   2,   we find
$x^2 - 4x + \left(\underline{\hspace{2cm}}\right) = -4$.   Completing the squares,
$(x - 2)^2 + \left(\underline{\hspace{2cm}}\right) = -4 + \frac{25}{16} +$ _____.   Therefore, the
center is _____ and the radius is _____.

OBJECTIVE C :   Find an equation of the circle passing through three
given noncollinear points.

8.   To find the circle determined by the three points $A(4,-1)$,
$B(-2,1)$,   and $C(6,5)$,   substitute the coordinates of each

---

4. $\sqrt{(y + 3)^2} = |y + 3|$, $\sqrt{x^2 + (y - 7)^2}$, $\sqrt{x^2 + (y - 7)^2} - 4$, $\sqrt{(y + 3)^2} + 4$,

$(y + 3)^2 + 8\sqrt{(y + 3)^2} + 16$, $8\sqrt{(y + 3)^2}$, $8(y + 3)$,   28y, $-8(y + 3)$,   12y - 48, $x^2 = 28y$

5. $(x - h)^2 + (y - k)^2 = r^2$ 6. $(x + 1)^2 + (y - 3)^2 = 11$,   -1

7. $y^2 + \frac{5y}{2}$, $\left(y + \frac{5}{4}\right)^2$,   4, $\left(2, -\frac{5}{4}\right)$, $\frac{5}{4}$

point into the equation $C_1x + C_2y + C_3 = -(x^2 + y^2)$. This gives the system of equations:

$$4C_1 - C_2 + C_3 = -(16 + 1)$$ from $A(4,-1)$,

_____   from $B(-2,1)$,

_____   from $C(6,5)$.

Solving by the method of elimination: subtraction of the second equation from the first gives _____ and the second from the third, _____. These last two equations may be solved for C_1 and C_2 as follows: 2 times the first added to the second gives $12C_1 + 8C_1 =$ _____ or $C_1 =$ _____. Substitution of this value of C_1 into the first equation $6C_1 - 2C_2 = -12$ gives $C_2 = -\frac{1}{2}(-12 +$ _____$)$ or $C_2 =$ _____. Substitution of these two values into the equation $4C_1 - C_2 + C_3 = -17$ gives $C_3 =$ _____. Therefore the required circle is _____.

OBJECTIVE D : Find an equation of a circle given three prescribed conditions on the circle (e.g., two points and a tangent at one of them).

9. Find the circle which is tangent to the line $2x + 3y = 12$ at $(3,2)$ and passes through the point $(6,-1)$.
Solution. The normal line to the circle at $(3,2)$ has slope _____, so an equation of the normal line is $y - 2 =$ _____, or $3x - 2y =$ _____. Given any chord of the circle, the line joining its midpoint to the center is perpendicular to the chord. In particular, the midpoint of the chord joining the two points $(3,2)$ and $(6,-1)$ is _____. Since the slope of the chord is $-3/3$, an equation of the perpendicular line joining the center $C(x,y)$ and midpoint is $y - \frac{1}{2} =$ _____ or $x - y =$ _____. Now, the center of the circle is located at the intersection of the two lines we have constructed: solving simultaneously,

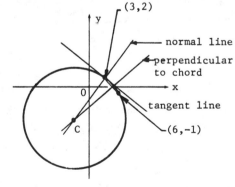

$$\left.\begin{array}{l} 3x - 2y = 5 \\ x - y = 4 \end{array}\right\}$$ yields $x = -3$ and $y =$ _____.

The radius r is the distance from the center to either one of the two given points, say $(3,2)$; thus,

_____

8. $-2C_1 + C_2 + C_3 = -(4 + 1)$, $6C_1 + 5C_2 + C_3 = -(36 + 25)$, $6C_1 - 2C_2 = -12$, $8C_1 + 4C_2 = -56$,

   $-24 - 56$, -4, 24, -6, -7, $x^2 + y^2 - 4x - 6y - 7 = 0$

$r = \sqrt{36 + \underline{\hspace{1.5cm}}} = \underline{\hspace{2cm}}.$ Therefore, the required circle is
$\underline{\hspace{5cm}}.$ A diagram illustrating our procedure is
given above.

8-3   PARABOLAS.

[OBJECTIVE A]:   Given the coordinates for the vertex V and the focus
F of a parabola, both of which lie along a line
parallel to a coordinate axis, find an equation of the
parabola and of its directrix. Sketch the graph
showing the focus, vertex, and directrix.

10.   Vertex at V(1,2) and focus at F(3,2)

The axis of symmetry of the parabola is the line containing
both V and F, or the line with equation $\underline{\hspace{2cm}}.$ Since
V is to the left of F on the axis of symmetry, the parabola
will open to the $\underline{\hspace{2cm}}.$ Thus, an equation of the
parabola has the form $\underline{\hspace{2cm}}.$
To calculate the number p, note
that it is the distance between the
vertex and the focus: thus,
p = $\underline{\hspace{1.5cm}}.$ Therefore, an equation
of the parabola is given by
$\underline{\hspace{4cm}}.$ The directrix
is given by $\underline{\hspace{2cm}}.$ Graph the
equation showing the focus, vertex,
and directrix.

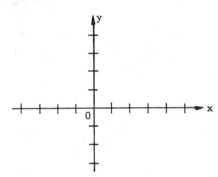

[OBJECTIVE B]:   Given the coordinates for the vertex V, and the
directrix L parallel to one of the coordinate axes,
find an equation of the parabola so determined and give
the coordinates of its focus. Sketch the graph showing
the focus, vertex, and directrix.

11.   Vertex at V(-1,2) and directrix L: y = 3

The axis of symmetry of the parabola is $\underline{\hspace{3cm}}$ to the
directrix. Since V is below the directrix, the parabola

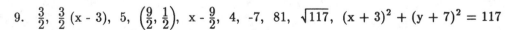

9. $\frac{3}{2}$, $\frac{3}{2}$ (x - 3), 5, $\left(\frac{9}{2}, \frac{1}{2}\right)$, x - $\frac{9}{2}$, 4, -7, 81, $\sqrt{117}$, $(x + 3)^2 + (y + 7)^2 = 117$

10. $y = 2$, right, $(y - k)^2 = 4p(x - h)$,

2, $(y - 2)^2 = 8(x - 1)$, x = -1

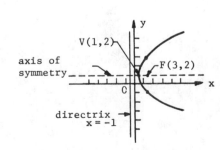

opens _____ and so an equation of the parabola has the form _____. The number p is the distance from the vertex to the directrix: thus, p = _____. Therefore an equation of the parabola is given by _____. The focus is located on the axis of symmetry p units from the vertex in the direction which the parabola opens. Thus the focus is _____. Graph the equation showing the focus, vertex, and directrix.

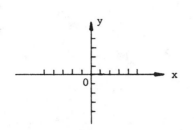

OBJECTIVE C : Given an equation of a parabola, find the vertex, axis of symmetry, focus, and directrix. Sketch the graph showing these features.

12. Consider the parabola $x^2 - 2x - 10y + 6 = 0$. Completing the square in the quadratic terms, _____ $- 10y + 6 = 1$ or _____ = 10 (_____). Therefore, the vertex of the parabola is _____. The axis of symmetry is _____ because the parabola opens _____. $4p =$ _____, so that p = _____. Therefore, the focus of the parabola is _____, and the directrix is given by _____. Sketch the graph showing these features.

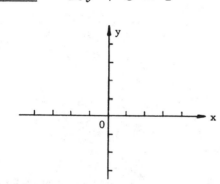

---

11. perpendicular, downward,

   $(x - h)^2 = -4p(y - k)$, 1,

   $(x + 1)^2 = -4(y - 2)$, F(-1,1)

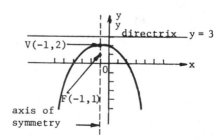

axis of symmetry

12. $x^2 - 2x + 1$, $(x - 1)^2$, $y - \frac{1}{2}$,

   $V\left(1,\frac{1}{2}\right)$, x = 1, upward, 10,

   $\frac{5}{2}$, F(1,3), y = -2

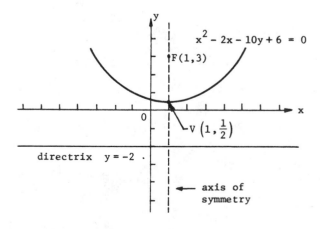

8-4 ELLIPSES.

13. In analyzing an ellipse, we find the intercepts are located on the axes of _____. The major axis is always in the direction of _____ axis length, and the foci always lie on the _____ axis.

14. If we use the letters a, b, and c to represent the lengths of the semimajor axis, semiminor axis, and half-distance between the foci, then it is always true that c^2 = _____.

15. The <u>eccentricity</u> of an ellipse is the ratio e = _____. It varies between _____ and measures the degree of departure of the ellipse from circularity: when e = 0, the ellipse is a _____, and when e = _____ it reduces to the line segment F_1F_2 joining the two foci.

[OBJECTIVE A]: Given an equation of an ellipse, find its center, vertices, and foci. Identify the major and minor axes, and sketch the graph showing all these features.

16. Consider the ellipse given by $9x^2 + y^2 - 18x + 2y + 9 = 0$. Thus, $9(x^2 - 2x) + ($_____$) = -9$, so completing the squares in each set of parentheses gives $9(x - 1)^2 +$ _____ $= 1$,

   or $\dfrac{(x - 1)^2}{1/9} +$ _____ $= 1$. Hence,

   $c^2 = a^2 - b^2 =$ _____, or
   c = _____. The center of the
   ellipse is _____ and the foci

   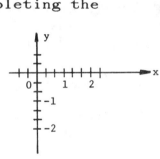

   are $F_1\left(1, -1 + \dfrac{2\sqrt{2}}{3}\right)$ and _____.

   The length of the semimajor axis is a = 1 and the length of the semimajor axis is b = _____. The vertices are the four points $(1, 0)$, $\left(\dfrac{2}{3}, -1\right)$, _____, and _____. The eccentricity of the ellipse is e = _____. Sketch the graph.

---

13. symmetry, largest, major   14. $a^2 - b^2$ 15. c/a, 0 and 1, circle, 1

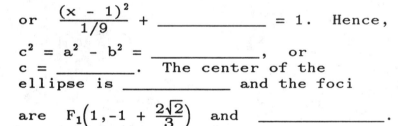

16. $y^2 + 2y$, $(y + 1)^2$, $\dfrac{(y + 1)^2}{1}$, $\dfrac{8}{9}$,

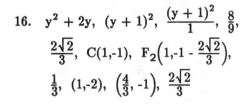

$\dfrac{2\sqrt{2}}{3}$, $C(1,-1)$, $F_2\left(1, -1 - \dfrac{2\sqrt{2}}{3}\right)$,

$\dfrac{1}{3}$, $(1,-2)$, $\left(\dfrac{4}{3}, -1\right)$, $\dfrac{2\sqrt{2}}{3}$

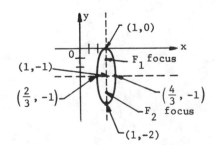

OBJECTIVE B :   Find an equation of an ellipse having a given center
C,   focus   F,   and semimajor axis   a.   Give the
eccentricity of the ellipse and sketch its graph.

17.   Center at $C(6,-1)$,   focus $F(6 + 3\sqrt{3}, -1)$, $a = 6$
The two foci are located along the major axis at a distance of
c units from the center of the ellipse. Thus,   c = _____.
Since $b^2 = a^2 - c^2$,   we find $b^2 = 36 -$ _____   so $b^2 =$ _____.
Thus, an equation of the ellipse is given by

_____.

The second focus of the ellipse
is _____. The
vertices are the points   (6,2),
(12,-1),   _____,   and
_____. The eccentricity is
e = _____. Sketch the
graph.

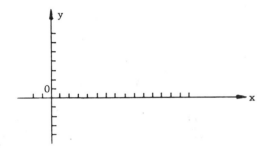

18.   To find an equation of the ellipse with center $C(-4,-2)$,
$c = \sqrt{7}$,   and   one vertex of the major axis at $(0,-2)$,   we note
first that the major axis is parallel to the _____ axis. The
value of a is the distance between the points $C(-4,-2)$ and
_____,   so   a = _____.   Therefore, $b^2 = a^2 - c^2 =$
_____ - 7 = _____.   Thus, an equation of the ellipse is given
by

_____.

The eccentricity is   e = _____.

8-5   HYPERBOLAS.

19.   The only differences between the equation of the ellipse and
the equation of the hyperbola are the _____ in the
equation of the hyperbola, and the new relation among   a,   b,
and   c;   namely, $a^2 - c^2 =$ _____.

---

17. $3\sqrt{3}$, 27, 9, $\dfrac{(x-6)^2}{36} + \dfrac{(y+1)^2}{9} = 1$,

$F_2(6 - 3\sqrt{3}, -1)$, (6,-4), (0,-1), $\dfrac{3\sqrt{3}}{6}$

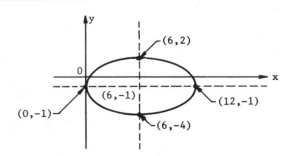

18.   x, (0,-2), 4, 16, 9, $\dfrac{(x+4)^2}{16} + \dfrac{(y+2)^2}{9} = 1$, $\dfrac{\sqrt{7}}{4}$ 19.   minus sign, $-b^2$

20.   There is no restriction   a > b   for the hyperbola as there is
       for the ellipse. The direction in which the hyperbola opens is
       controlled by the _____ rather than by the relative
       _____ of the coefficients of the quadratic terms.

21.   To obtain the <u>asymptotes</u> for a hyperbola, set the expression
       _____ or _____ equal to zero (depending on
       how the hyperbola is defined), and then factor and solve the
       resulting quadratic equation.

22.   The <u>eccentricity</u> for a hyperbola is defined as   e = _____,
       and <u>is always</u> _____ one.

[OBJECTIVE]:   Given an equation representing a hyperbola, find the
               center, vertices, foci, and asymptotes. Sketch a graph
               showing all these features.

23.   Consider the hyperbola given by $\dfrac{(x - 2)^2}{4} - \dfrac{(y + 1)^2}{5} = 1$.
       Because of the minus sign associated with the   y   terms, the
       hyperbola opens _____ and _____. The center of the
       hyperbola is   C(h,k) = _____.
       Now, $a^2 =$ _____ and $b^2 =$ _____,
       so $c^2 = a^2 + b^2 =$ _____. The
       coordinates of the foci are
       $F_1(h-c,k)$ and $F_2(h+c,k)$ or
       _____ and _____. The
       coordinates of the vertices are
       $V_1(h-a,k)$ and $V_2(h+a,k)$ or
       _____ and _____. Equations
       may be found for the asymptotes by
       setting $\dfrac{(x - 2)^2}{4} - \dfrac{(y + 1)^2}{5}$ equal
       to _____. Thus, equations of the asymptotes are
       _____. The eccentricity is   e = _____.
       Sketch the graph of the hyperbola.

---

20. signs, sizes   21. $\dfrac{x^2}{a^2} - \dfrac{y^2}{b^2}, \dfrac{y^2}{a^2} - \dfrac{x^2}{b^2}$ 22. c/a, greater than or equal to

23. right, left, C(2,-1), 4, 5, 9,

     $F_1(-1,-1)$, $F_2(5,-1)$, $V_1(0,-1)$

     $V_2(4,-1)$, zero, $y + 1 = \pm \dfrac{\sqrt{5}}{2} (x - 2)$,

     $\dfrac{c}{a} = \dfrac{3}{2}$

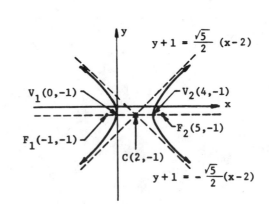

24. Consider the hyperbola $9x^2 - 4y^2 + 18x + 16y + 29 = 0$. Thus, $9(x^2 + 2x) - 4(\underline{\hspace{3cm}}) = -29$. Completing the squares in each set of parentheses gives $9(x + 1)^2 = -4(\underline{\hspace{2cm}}) = -29 + 9 - \underline{\hspace{1.5cm}} = \underline{\hspace{1.5cm}}$, or $\dfrac{(y - 2)^2}{9} - \underline{\hspace{2cm}} = 1$. Because of the minus sign, the hyperbola opens $\underline{\hspace{1.5cm}}$ and $\underline{\hspace{1cm}}$. The center of the hyperbola is $C(h,k) = \underline{\hspace{2cm}}$. The coordinates of the vertices are $V_1(h,k-a)$ and $V_2(h,k+a)$ or $\underline{\hspace{2cm}}$ and $\underline{\hspace{2cm}}$. Now, $c^2 = a^2 + b^2 = \underline{\hspace{1.5cm}}$, and the coordinates of the foci are $F_1(h,k-c)$ and $F_2(h,k+c)$ or $\underline{\hspace{3cm}}$ and $\underline{\hspace{2cm}}$. Equations for the asymptotes may be found by setting the expression $\dfrac{(y - 2)^2}{9} - \dfrac{(x + 1)^2}{4}$ equal to $\underline{\hspace{2.5cm}}$. Thus, equations of the asymptotes are $\underline{\hspace{4cm}}$. The eccentricity is $e = \underline{\hspace{2cm}}$. Sketch the graph showing the features of the hyperbola.

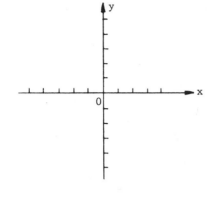

8-6   THE GRAPHS OF QUADRATIC EQUATIONS.

25. Any second degree equation in x and y represents a circle, parabola, ellipse, or hyperbola (although it may degenerate). To find the curve given its equation
$$Ax^2 + Bxy + Cy^2 + Dx + Ey + F = 0,$$

   (1) First rotate axes, to force $B = 0$, through an angle α satisfying $\cot 2\alpha = \underline{\hspace{3cm}}$.

   (2) Next, $\underline{\hspace{3cm}}$ axes by completing the squares (if necessary) to reduce the equation to a standard form.

---

24. $y^2 - 4y$, $(y - 2)^2$, 16, -36, $\dfrac{(x + 1)^2}{4}$,

   up, down, $C(-1,2)$, $V_1(-1,-1)$, $V_2(-1,5)$,

   13, $F_1(-1, 2 - \sqrt{13})$, $F_2(-1, 2 + \sqrt{13})$,

   zero, $y - 2 = \pm \dfrac{3}{2}(x + 1)$, $\dfrac{c}{a} = \dfrac{\sqrt{13}}{3}$

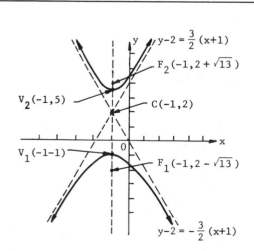

25. $\dfrac{A - C}{B}$, translate

OBJECTIVE : Given an equation of the form $Ax^2 + Bxy + Cy^2 + F = 0$, transform the equation by a rotation of axes into an equation that has no cross-product term.

26. Consider the equation $5x^2 - 2\sqrt{3}xy + 7y^2 = 6$. The required angle of rotation to eliminate the xy term satisfies

$\cot 2\alpha = \dfrac{5 - 7}{\underline{\hspace{1cm}}} = \underline{\hspace{1.5cm}}$; hence $2\alpha = \underline{\hspace{1.5cm}}$ radians, or

$\alpha = \underline{\hspace{1.5cm}}$. Thus, $\sin \alpha = \underline{\hspace{1.5cm}}$ and $\cos \alpha = \underline{\hspace{1.5cm}}$. The

equations for the rotation of axes are $x = \dfrac{\sqrt{3}}{2}x' - \dfrac{1}{2}y'$ and

$\underline{\hspace{5cm}}$. Substitution for x and y into the given second-degree equation gives

$$5\left(\tfrac{3}{4}x'^2 - \underline{\hspace{2cm}} + \tfrac{1}{4}y'^2\right) - 2\sqrt{3}\left(\tfrac{\sqrt{3}}{4}x'^2 + \underline{\hspace{2cm}} - \tfrac{\sqrt{3}}{4}y'^2\right)$$
$$+ 7\left(\tfrac{1}{4}x'^2 + \underline{\hspace{2cm}} + \tfrac{3}{4}y'^2\right) = 6$$

Simplifying algebraically,
$$\left(\tfrac{15}{4} - \underline{\hspace{1.5cm}} + \tfrac{7}{4}\right)x'^2 + \left(\underline{\hspace{1cm}} + \tfrac{3}{2} + \tfrac{21}{4}\right)y'^2 = 6, \quad \text{or}$$

$\underline{\hspace{4cm}}$. This is an equation of $\underline{\hspace{3cm}}$.

27. Consider the equation $3x^2 - 2xy + y^2 = 8$. The angle of rotation for elimination of the cross-product term satisfies

$\cot 2\alpha = \dfrac{\underline{\hspace{1.5cm}}}{-2} = \underline{\hspace{1.5cm}}$; thus, $2\alpha = \underline{\hspace{1.5cm}}$ radians. It

follows that $\cos 2\alpha = \underline{\hspace{1.5cm}}$. Thus, using the trigonometric half-angle formulas we find

$\sin \alpha = \sqrt{\dfrac{1 - \cos 2\alpha}{2}} = \underline{\hspace{3cm}}$, and

$\cos \alpha = \sqrt{\dfrac{1 + \cos 2\alpha}{2}} = \underline{\hspace{3cm}}$.

Therefore, the rotational equations are
$x = \tfrac{1}{2}\sqrt{2 - \sqrt{2}}\ x' - \tfrac{1}{2}\sqrt{2 + \sqrt{2}}\ y'$ and

$\underline{\hspace{6cm}}$. Substitution for x and y into the original second degree equation gives,

$\underline{\hspace{5cm}}$.

This equation represents $\underline{\hspace{3cm}}$.

---

26. $-2\sqrt{3}$, $\dfrac{1}{\sqrt{3}}$, $\dfrac{\pi}{3}$, $\dfrac{\pi}{6}$, $\dfrac{1}{2}$, $\dfrac{\sqrt{3}}{2}$, $y = \tfrac{1}{2}x' + \tfrac{\sqrt{3}}{2}y'$, $\dfrac{\sqrt{3}}{2}x'y'$, $\tfrac{1}{2}x'y'$, $\dfrac{\sqrt{3}}{2}x'y'$, $\dfrac{3}{2}$,

$\dfrac{5}{4}$, $4x'^2 + 8y'^2 = 6$, an ellipse

27. $3 - 1$, -1, $\dfrac{3\pi}{4}$, $-\dfrac{\sqrt{2}}{2}$, $\tfrac{1}{2}\sqrt{2 + \sqrt{2}}$, $\tfrac{1}{2}\sqrt{2 - \sqrt{2}}$, $y = \tfrac{1}{2}\sqrt{2 + \sqrt{2}}\ x' + \tfrac{1}{2}\sqrt{2 - \sqrt{2}}\ y'$,

$\left(2 - \sqrt{2}\right)x'^2 + \left(2 + \sqrt{2}\right)y'^2 = 8$, an ellipse

8-7 PARABOLA, ELLIPSE, OR HYPERBOLA?   THE DISCRIMINANT TELLS.

OBJECTIVE A : Given a second degree equation of the form $Ax^2 + Bxy + Cy^2 + Dx + Ey + F = 0$, use the discriminant to classify it as representing a circle, an ellipse, a parabola, or a hyperbola.

28. The <u>discriminant</u> is the invariant expression _____. Both the discriminant and the expression _____ are invariant under rotations of axes.

29. If the discriminant is positive, the equation represents _____.

30. If the discriminant is zero, the equation represents _____.

31. If the discriminant is negative, the equation represents _____.

32. In order that the equation represent a circle, it is necessary that the discriminant be _____ and that _____.

33. Consider the equation given by
$$2x^2 - 4xy - y^2 + 20x - 2y + 17 = 0.$$
The discriminant is $B^2 - 4AC = 16 - ($_____$) = $_____. Thus, the equation represents _____.

34. For $4x^2 - 12xy + 9y^2 - 52x + 26y + 81 = 0$ the discriminant is $B^2 - 4AC = $_____ $- 4(36) = $_____. Thus, the equation represents _____.

OBJECTIVE B : Use the invariants $B^2 - 4AC$ and $A + C$ to determine an equation to which $Ax^2 + Bxy + Cy^2 + F = 0$ reduces when the axes are rotated to eliminate the cross-product term.

35. Consider again the equation $3x^2 - 2xy + y^2 = 8$, as in Problem 27 above. Now, $B^2 - 4AC = $_____ and $A + C = $_____. Thus, $A' + C' = $_____ and $-4A'C' = $_____ or $C' = $_____. Substitution into the first equation $A' + C' = 4$ gives $4 = A' + C' = A' + $_____ or $A'^2 - 4A' + 2 = 0$. Solving this quadratic equation for A' gives $A' = $_____ or $A' = $_____; thus, $C' = 4 - A'$ is given by $C' = $_____ or $C' = $_____. Thus, we find an equation _____ or _____ with no cross-product term.

---

28. $B^2 - 4AC$, $A + C$ 29. a hyperbola   30. a parabola   31. an ellipse

32. negative, $A = C$ 33. -8, 24, a hyperbola   34. 144, 0, a parabola

35. -8, 4, 4, -8, $\dfrac{2}{A'}$, $\dfrac{2}{A'}$, $2 + \sqrt{2}$, $2 - \sqrt{2}$, $2 - \sqrt{2}$, $2 + \sqrt{2}$, $\left(2 + \sqrt{2}\right) x'^2 + \left(2 - \sqrt{2}\right) y'^2 = 8$
or $\left(2 - \sqrt{2}\right) x'^2 + \left(2 + \sqrt{2}\right) y'^2 = 8$

8-8   SECTIONS OF A CONE.

36.   The circle, parabola, ellipse, and hyperbola are known as
_____ because each may be obtained by cutting a
_____ by a _____ .

37.   The type of section is determined by the relationship between
the angle α the cutting plane makes with the _____
and the angle β which _____ the cone.

38.   The section is

_____ ,   if $\alpha = \beta$;

_____ ,   if $0 \leq \alpha < \beta$;

_____ ,   if $\alpha = 90°$;

_____ ,   if $\beta < \alpha < 90°$.

8-9   PARAMETRIC EQUATIONS FOR CONICS AND OTHER CURVES.

[OBJECTIVE A]:   Sketch the graph of a curve given in parametric form
$x = f(t)$ and $y = g(t)$ as the parameter t varies
over a given domain. Also, find a cartesian equation
for the curve.

39.   Consider the curve given by the parametric equations
$x = t - 2$, $y = 2t + 3$, $-\infty < t < \infty$. Complete the following
table providing some of the points $P(x,y)$ on the curve:

t	-2	-1	0	1	2	3
x	-4					
y	-1					

To eliminate the parameter t,
note that $t = x + 2$.
Substitution for t in the
parametric equation for y
gives $y = $ _____ . This is
a cartesian equation for a
_____ with slope $m = $ _____
and y-intercept $b = $ _____ .
Sketch the curve in the coordinae system to the right.

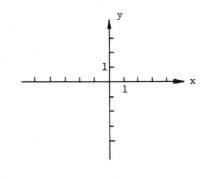

---

36.   conic sections, cone, plane   37.   axis of the cone, generates

38.   a parabola, a hyperbola, a circle, an ellipse

39.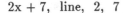

t	-2	-1	0	1	2	3
x	-4	-3	-2	-1	0	1
y	-1	1	3	5	7	9

$2x + 7$, line, 2, 7

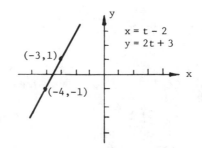

40. Consider the curve $x = a\cos^3 t$, $y = a\sin^3 t$,
     $-\infty < t < +\infty$. Complete the following table providing some of
     the points $P(x,y)$ on the curve:

t	0	$\frac{\pi}{6}$	$\frac{\pi}{4}$	$\frac{\pi}{3}$	$\frac{\pi}{2}$	$\frac{3\pi}{4}$	π	$\frac{3\pi}{2}$	2π
x	a								
y	0								

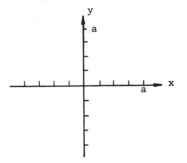

To eliminate the parameter t, observe
that $\left(\frac{x}{a}\right)^{1/3} = $ _____ and

$\left(\frac{y}{a}\right)^{1/3} = $ _____.   Thus,

$\left(\frac{x}{a}\right)^{1/3} + \left(\frac{y}{a}\right)^{2/3} = $ _____ = _____,

or $x^{2/3} + y^{2/3} = $ _____.   This curve

is a <u>hypocycloid</u>. Sketch its graph and notice its symmetries.

41. For the curve given by the parametric
     equations $x = e^t$ and $y = e^{-t}$,
     $-\infty < t < \infty$, complete the following
     table:

t	-2	-1	0	1	2	3
x						
y						

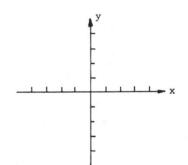

To eliminate the paraemter t, notice
that $xy = $ _____.   This equation
describes a _____.   Sketch the
graph in the coordinate system at the
right.

40.

t	0	$\frac{\pi}{6}$	$\frac{\pi}{4}$	$\frac{\pi}{3}$	$\frac{\pi}{2}$	$\frac{3\pi}{4}$	π	$\frac{3\pi}{2}$	2π
x	a	$\approx.6a$	$\frac{a}{2\sqrt{2}}$	$\frac{a}{8}$	0	$-\frac{a}{2\sqrt{2}}$	-a	0	a
y	0	$\frac{a}{8}$	$\frac{a}{2\sqrt{2}}$	$\approx.6a$	a	$\frac{a}{2\sqrt{2}}$	0	-a	0

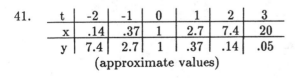

$\cos t$, $\sin t$, $\cos^2 t + \sin^2 t$, 1, $a^{2/3}$

41.

t	-2	-1	0	1	2	3
x	.14	.37	1	2.7	7.4	20
y	7.4	2.7	1	.37	.14	.05

(approximate values)

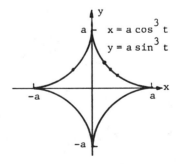

$xy = 1$, hyperbola

42.   For the curve given parametrically in Problem 41, notice that
x and y are always positive. Are the parametric equations
and the cartesian equation coextensive? _____ , because x
and y can both be _____ in the cartesian equation
xy = 1.

OBJECTIVE B :   Find parametric equations for a curve described
geometrically, or by an equation, in terms of some
specified or arbitrary parameter.

43.   Find parametric equations for the circle with center C(-2,3)
and radius $r = \sqrt{2}$.

Solution.   An equation of the circle is
$(x + 2)^2 +$ _____ $=$ _____ ,   or _____ $= 1$.

This suggests the substitutions $\dfrac{x + 2}{\sqrt{2}} = \sin\theta$ and $\dfrac{y - 3}{\sqrt{2}} =$

_____.   Hence, parametric equations for the circle are
x = _____ and y = _____ , $0 \le \theta \le 2\pi$.

44.   Find parametric equations for the line in the plane through the
point (a,b) with slope m,   where the parameter t   is the
change x - a.

Solution.   For any point P(x,y)   on the line,
y - b = m(_____) = _____.   Thus,   x = _____ and y = _____
give parametric equations of the line in terms of the specified
parameter t.

---

42. No, negative

43. $(y - 3)^2$, 2, $\left(\dfrac{x + 2}{\sqrt{2}}\right)^2 + \left(\dfrac{y - 3}{\sqrt{2}}\right)^2$, $\cos\theta$, $\sqrt{2}\sin\theta - 2$, $3 + \sqrt{2}\cos\theta$

44. x - a, mt, a + t, b + mt

CHAPTER 8   SELF-TEST

1. Find an equation for the points $P(x,y)$ whose distances from the point $(-4,2)$ are always twice their distance from the point $(6,-2)$.

2. Write an equation for the circle with center $C(4,6)$ and radius $r = 3$.

3. Find an equation of the circle with center at $C(1,2)$ and tangent to the line $x + 2y = 10$.

4. Find an equation of the circle passing through the three points $A(4,2)$, $B(6,0)$, and $C(0,-6)$.

5. Find an equation of the circle inscribed in the triangle formed by the lines $y = 0$, $x = 0$, and $x + y = 8$.

6. Find an equation of the parabola with vertex $V(2,0)$ and focus the origin. Sketch the graph.

7. Find an equation of the parabola whose directrix is the line $y = -5$ with vertex $V(4,-3)$. Sketch.

8. Find the vertex, the focus, and the directrix of the parabola $x^2 - 4x + 6y + 34 = 0$, and sketch the graph.

9. Find the center, vertices, foci, and eccentricity of the ellipse $9x^2 + 16y^2 - 54x + 128y + 193 = 0$, and sketch the graph.

10. Write an equation of the ellipse with center $C(2,5)$, focus $F(-1,5)$, and semiminor axis $= 4$.

11. Find the center, vertices, foci, and asymptotes of the hyperbola $9x^2 - 4y^2 - 18x + 32y - 91 = 0$, and sketch the graph.

12. Consider the equation $x^2 - 2xy - y^2 = 12$.
    (a) Use the discriminant to classify it.
    (b) Transform the equation by a rotation of axes into an equation with no cross-product term.

13. Use the invariants $B^2 - 4AC$ and $A + C$ to determine an equation to which $x^2 + xy + y^2 = 10$ reduces when the axes are rotated to eliminate the cross-product term.

14. Find parametric equations for the curve described by the point $P(x,y)$ for $t \geq 0$ if its coordinates satisfy

$$\frac{dx}{dt} = \sqrt{y}, \quad \frac{dy}{dt} = 2y; \quad t = 0, \quad x = -5, \quad y = 4.$$

15. Sketch the graph of the curve described by $x = t - 2$ and $y = t^2 - t + 1$ for $-\infty < t < \infty$. Also find a cartesian equation of the curve.

16. Find parametric equations for the circle $x^2 + y^2 = 2x$,   using
     as parameter the arc length   s   measured counterclockwise from
     the point   (2,0)   to the point   (x,y).

SOLUTIONS TO CHAPTER 8 SELF-TEST

1. Let d_1 denote the distance from   P   to   (-4,2)   and d_2 the
    distance from   P   to   (6,-2).   Then $d_1 = 2d_2$,   or $d_1^2 = 4d_2^2$.
    Therefore, $(x + 4)^2 + (y - 2)^2 = 4[(x - 6)^2 + (y + 2)^2]$,   or
    expanding algebraically and collecting like terms,
    $3x^2 - 56x + 3y^2 + 20y + 140 = 0$.   Completing the squares in the
    x   and   y   terms, $3\left(x - \frac{28}{3}\right)^2 + 3\left(y + \frac{10}{3}\right)^2 = \frac{16 \cdot 29}{3}$.   This
    represents a circle of radius $r = \frac{4}{3}\sqrt{29}$ with center
    $C\left(\frac{28}{3}, \frac{-10}{3}\right)$.

2. $(x - 4)^2 + (y - 6)^2 = 9$,   or $x^2 + y^2 - 8x - 12y + 43 = 0$.

3. The distance between the point   (h,k)   and the line $ax + by = c$
    is
    $$\frac{|ah + bk - c|}{\sqrt{a^2 + b^2}}.$$
    Now, the radius of the circle is the distance from its center to
    the tangent line, and we find this to be
    $$r = \frac{|1 \cdot 1 + 2 \cdot 2 - 10|}{\sqrt{1 + 4}} = \sqrt{5}$$

    Thus, an equation of the circle is $(x - 1)^2 + (y - 2)^2 = 5$.

4. Using the equation $C_1 x + C_2 y + C_3 = -(x^2 + y^2)$ for a circle,
    by substitutions we obtain the equations,

     $4C_1 + 2C_2 + C_3 = -20$ from point   A,

     $6C_1 \qquad + C_3 = -36$ from point   B,

     $\qquad - 6C_2 + C_3 = -36$ from point   C.
    We solve these by elimination: $-\frac{6}{4}$ times the first equation
    added to the second equation gives
    $$-3C_2 - \frac{1}{2}C_3 = -6, \quad \text{or} \quad -6C_2 - C_3 = -12.$$
    Addition of this last equation and the third equation results in
    $$-12C_2 = -48 \quad \text{or} \quad C_2 = 4.$$
    From the third equation, $C_3 = -36 + 6C_2$,   and substitution of the
    value $C_2 = 4$ then gives $C_3 = -12$.
    From the second equation, $6C_1 = -36 - C_3$,   and substitution of
    $C_3 = -12$ gives $6C_1 = -24$,   or $C_1 = -4$.
    Thus, an equation for the circle is $x^2 + y^2 - 4x + 4y - 12 = 0$.
    Completing the squares, $(x - 2)^2 + (y + 2)^2 = 20$.

5. Since the circle is tangent to the line x + y = 8 with slope
    m = -1, the center of the circle must lie on the line y = x.
    Let C(h,h) denote the center. Then the radius of the circle is
    r = h because the circle is tangent to each coordinate axis.
    However, r is also equal to the distance from the center to the
    line x + y = 8. Thus,

    $$h = \frac{|1 \cdot h + 1 \cdot h - 8|}{\sqrt{1 + 1}} = \frac{|2h - 8|}{\sqrt{2}}$$

    Squaring both sides and simplifying, we obtain,

    $$h^2 - 16h + 32 = 0.$$

    Solution of this quadratic equation for $0 \le h \le 8$ gives,
    $h = 8 - 4\sqrt{2} \approx 2.343$ units. An equation for the circle is
    (approximately) given by

    $$(x - 2.343)^2 + (y - 2.343)^2 = 5.49.$$

6. The parabola opens to the left and
    hence has an equation of the form
    $(y - k)^2 = -4p(x - h)$. The point
    (h,k) is the vertex V(2,0).
    The value of p is the distance
    from the vertex to the focus, so
    p = 2. Thus, an equation of the
    parabola is $y^2 = -8(x - 2)$. The
    directrix is the line x = 4.

    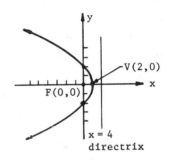

7. Since the directrix is parallel to
    the x-axis and the vertex lies
    above it (i.e., -3 > -5), the
    parabola opens upward and hence
    has an equation of the form
    $(x - h)^2 = 4p(y - k)$. The value
    of p is the distance from the
    vertex to the directrix, so p = 2.
    Thus, an equation of the parabola
    is $(x - 4)^2 = 8(y + 3)$. The
    focus is F(4,-1).

    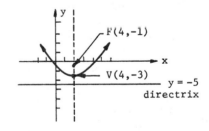

8. Completing the square in x,
    $(x - 2)^2 = -6(y + 5)$. The
    parabola opens downward with
    4p = 6 or p = 3/2. The vertex
    is located at V(2,-5) and the
    focus is F(2,-5-p) = F(2,-13/2).
    The directrix is y = -5 + p or
    y = - 7/2. The parabola is sketched
    at the right.

    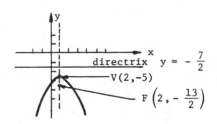

9. Completing the squares in the x
    and y terms,

    $9(x^2 - 6x + 9) + 16(y^2 + 8y + 16)$
    $= 81 + 256 - 193,$

    which may be written as

    $9(x - 3)^2 + 16(y + 4)^2 = 144,$

    or

    $\dfrac{(x - 3)^2}{16} + \dfrac{(y + 4)^2}{9} = 1.$

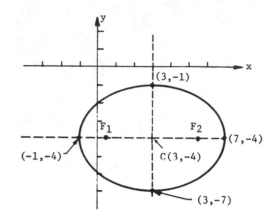

    The center is C(3,-4). The semi-
    major axis is a = 4, and the
    semiminor axis is b = 3. The foci
    lie on the line y = -4. Now
    $c^2 = a^2 - b^2 = 16 - 9 = 7.$ Thus,
    the foci are located at $F_1(3-\sqrt{7},-4)$
    and $F_2(3+\sqrt{7},-4)$. The vertices are
    $V_1(-1,-4)$, $V_2(7,-4)$, $V_3(3,-1)$,   and $V_4(3,-7)$. The graph is
    shown at the right. The eccentricity is $e = c/a = \sqrt{7}/4 \approx 0.661.$

10. The focus lies on the major axis at a distance   c = 2 - (-1) = 3
    units from the center   C(2,5). The major axis is along the line
    y = 5 which contains the center and focus. Now, $a^2 = b^2 + c^2 =$
    16 + 9 = 25. Hence, an equation of the ellipse is

    $$\frac{(x - 2)^2}{25} + \frac{(y - 5)^2}{16} = 1 \quad \text{or} \quad 16(x - 2)^2 + 25(y - 5)^2 = 400.$$

11. Completing the squares in the x
    and y terms,

    $9(x^2 - 2x + 1) - 4(y^2 - 8y + 16) = 36,$

    or

    $\dfrac{(x - 1)^2}{4} - \dfrac{(y - 4)^2}{9} = 1.$

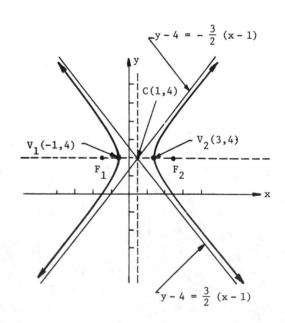

    The center is C(1,4). Now,
    $c^2 = a^2 + b^2 = 4 + 9 = 13.$ Since
    b > a, the hyperbola opens to the
    left and to the right. Thus, the
    foci are $F_1(1-\sqrt{13},4)$ and
    $F_2(1+\sqrt{13},4)$. The vertices are
    found by setting y = 4; whence
    x - 1 = ±2. We find $V_1(-1,4)$ and
    $V_2(3,4)$. Finally, the asymptotes
    are obtained by setting

    $\dfrac{(x - 1)^2}{4} + \dfrac{(y - 4)^2}{9}$ equal to zero,

    so $y - 4 = \pm\frac{3}{2}(x - 1)$. The graph
    is sketched at the right.

12. (a) The discriminant is $B^2 - 4AC = (-2)^2 - 4(1)(-1) = 8 > 0$. Therefore, the equation represents a hyperbola.

(b) $\cot 2\alpha = \dfrac{A - C}{B} = \dfrac{1 + 1}{-2} = -1$; thus $2\alpha = \dfrac{3\pi}{4}$ radians. Hence, $\cos 2\alpha = -1/\sqrt{2}$. Using the half-angle formulas,

$$\sin \alpha = \sqrt{\frac{1 - \cos 2\alpha}{2}} = \sqrt{\frac{1 + \sqrt{2}}{2\sqrt{2}}} = \frac{1}{2} \sqrt{2 + \sqrt{2}}, \quad \text{and}$$

$$\cos \alpha = \sqrt{\frac{1 + \cos 2\alpha}{2}} = \sqrt{\frac{1 - \sqrt{2}}{2\sqrt{2}}} = \frac{1}{2} \sqrt{2 - \sqrt{2}}.$$

Now, $A' = A \cos^2 \alpha + B \cos \alpha \sin \alpha + C \sin^2 \alpha$
$$= \tfrac{1}{4}(2 - \sqrt{2}) - \tfrac{2}{4}(\sqrt{-2 + 4}) - \tfrac{1}{4}(2 + \sqrt{2}) = -\sqrt{2}$$

$C' = A \sin^2 \alpha - B \cos \alpha \sin \alpha + C \cos^2 \alpha$
$$= \tfrac{1}{4}(2 + \sqrt{2}) + \tfrac{2}{4}(\sqrt{-2 + 4}) - \tfrac{1}{4}(2 - \sqrt{2}) = \sqrt{2}.$$

Thus, since $B' = 0$ because of the choice $\alpha = \dfrac{3\pi}{8}$, the original equation is reduced to $A'x'^2 + C'y'^2 = F$, or $-\sqrt{2}\, x'^2 + \sqrt{2}\, y'^2 = 12$, or $y'^2 - x'^2 = 6\sqrt{2}$.

13. Now, $B^2 - 4AC = 1 - 4(1)(1) = -3 = -4A'C'$; thus $A'C' = \dfrac{3}{4}$. Also, $A + C = 2 = A' + C'$. Therefore, $\dfrac{3}{4} = A'(2 - A') = 2A' - A'^2$, or $4A'^2 - 8A' + 3 = 0$. Solving this quadratic equation,

$$A' = \frac{8 \pm \sqrt{64 - 48}}{8}. \quad \text{Thus,} \quad A' = \frac{3}{2} \quad \text{or} \quad A' = \frac{1}{2}.$$

For $A' = \dfrac{3}{2}$, $C' = \dfrac{1}{2}$ and the original equation reduces to

$$\tfrac{3}{2} x'^2 + \tfrac{1}{2} y'^2 = 10 \quad \text{or} \quad 3x'^2 + y'^2 = 20.$$

On the other hand, if $A' = \dfrac{1}{2}$, $C' = \dfrac{3}{2}$ and the original equation reduces to

$$x'^2 + 3y'^2 = 20.$$

We recognize either of these equations as representing an ellipse.

14. From $\dfrac{dy}{dt} = 2y$ we obtain $\dfrac{dy}{y} = 2\, dt$. Thus, $\ln y = 2t + \ln C_1$ or $y = C_1 e^{2t}$. From the initial condition $t = 0$ and $y = 4$ we find $C_1 = 4$ so that $y = 4e^{2t}$. Next, $\dfrac{dx}{dt} = \sqrt{y}$ implies

$dx = 2e^t\, dt$, or $x = 2e^t + C_2$. From the initial condition $t = 0$ and $x = -5$ we find $C_2 = -7$; hence $x = 2e^t - 7$.

15.   We have the following table
       giving some of the points on
       the curve:

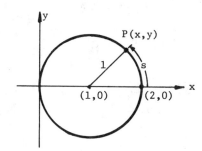

$$x = t - 2$$
$$y = t^2 - t + 1$$

t	-1	0	1	2	3
x	-3	-2	-1	0	1
y	3	1	1	3	7

Substitution of $t = x + 2$
into the parametric equation
for y gives
$y = (x + 2)^2 - (x + 2) + 1$,   or simplifying algebraically,
$y = x^2 + 3x + 3$.   This is a cartesian equation of the parabola
sketched in the figure above.

16.   Completing the square gives the
       equation $(x - 1)^2 + y^2 = 1$
       which we recognize as an equation
       of a unit circle centered at
       $(1,0)$ (see the figure at the
       right). Hence, $x - 1 = 1 \cos s$
       and $y - 0 = 1 \sin s$ or,

         $x = 1 + \cos s$ and $y = \sin s$

       are parametric equations for
       the circle in terms of the arc
       length parameter s.

CHAPTER 9 HYPERBOLIC FUNCTIONS

INTRODUCTION.

   In this chapter you will study the hyperbolic functions and see
that the identities and differentiation formulas associated with them
bear a striking resemblance to those associated with the trigonometric
functions.   Also, there is a direct connection between the inverse
hyperbolic functions and the natural logarithm.   These functions are
useful in applied mathematics, in certain integration methods, and in
solving differential equations.   They arise in the study of falling
bodies, hanging cables, ocean waves, and other phenomena in science
and engineering.

9-1 DEFINITIONS AND IDENTITIES.

   ☐ OBJECTIVE A ☐:   Define the six hyperbolic functions and graph them.

1.   cosh x = _____.
     Sketch the graph at the right.

2.   sinh x = _____.
     Sketch the graph at the right
     on the same coordinate system
     as the cosh x.

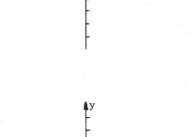

3.   tanh x = _____.

4.   coth x = _____.
     Sketch the graphs at the right.

---

1. $\dfrac{e^x + e^{-x}}{2}$ 　　2. $\dfrac{e^x - e^{-x}}{2}$ 　　3. $\dfrac{\sinh x}{\cosh x}$ 　　4. $\dfrac{\cosh x}{\sinh x}$

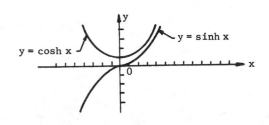

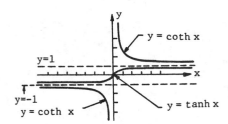

5.   sech x = _____ .
     Sketch the graph at the right.

6.   csch x = _____ .

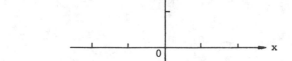

OBJECTIVE B :   Given the value for one of the six hyperbolic functions
                 at a point, determine the values of the remaining five
                 at that point.

7.   Suppose   tanh x = $-\frac{\sqrt{3}}{2}$.   Then,

     $\text{sech}^2 x = 1 - $ _____ = _____ or sech x = _____ .

     Thus,   cosh x = $\frac{1}{\text{sech } x}$ = _____ .

8.   Continuing Problem 7, $\sinh^2 x = \cosh^2 x - $ _____ = _____ .
     Since   tanh x   is negative and   cosh x   is positive, it follows
     that   sinh x = _____ .   Then,   csch x = _____ and
     coth x = _____ .

9.   cosh (-x) = _____ .

10. sinh (-x) = _____ .

11. sinh (x + y) = _____ .

12. cosh (x + y) = _____ .

13. sinh 2x = _____ .

14. cosh 2x = _____ .

15. cosh 2x - 1 = _____ .

---

5. $\dfrac{1}{\cosh x}$ 6. $\dfrac{1}{\sinh x}$

7. $\tanh^2 x$, $\frac{1}{4}$, $\frac{1}{2}$, 2

8. 1, 3, $-\sqrt{3}$, $-\dfrac{1}{\sqrt{3}}$, $\dfrac{-2}{\sqrt{3}}$

9. cosh x   10. -sinh x

11. sinh x cosh y + cosh x sinh y

12. cosh x cosh y + sinh x sinh y

13. 2 sinh x cosh x

14. $\cosh^2 x + \sinh^2 x$

15. $2 \sinh^2 x$

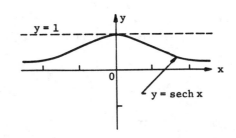

9-2 DERIVATIVES AND INTEGRALS.

OBJECTIVE A : Calculate the derivatives of functions expressed in terms of hyperbolic functions.

16. $\frac{d}{dx} \cosh u = $ _____ .

17. $\frac{d}{dx} \sinh u = $ _____ .

18. $\frac{d}{dx} \tanh u = $ _____ .

19. $y = \sinh^3 (3 - 2x^2)$

$\frac{dy}{dx} = 3 \sinh^2 (3 - 2x^2) \cdot \frac{d}{dx}$ _____

$= 3 \sinh^2 (3 - 2x^2) \cosh (3 - 2x^2) \frac{d}{dx}$ _____

$=$ _____ .

20. $y = e^x \tanh 2x$

$\frac{dy}{dx} = e^x \frac{d}{dx} ($ _____ $) + e^x \tanh 2x$

$= e^x$ _____ $\frac{d}{dx} ($ _____ $) + e^x \tanh 2x$

$=$ _____ .

21. $y = x^{\sinh x}, \quad x > 0$

$\frac{dy}{dx} = \frac{d}{dx} (e^{\sinh x \cdot \ln x}) = e^{\sinh x \cdot \ln x} \frac{d}{dx} ($ _____ $)$

$= e^{\sinh x \cdot \ln x} (\sinh x \cdot \frac{d}{dx} \ln x + \ln x \cdot \frac{d}{dx}$ _____ $)$

$= e^{\sinh x \cdot \ln x} ($ _____ $)$

$= x^{\sinh x - 1} ($ _____ $)$.

22. $e^y = \text{sech } x$

Differentiating implicitly, $\frac{d}{dx}(e^y) = \frac{d}{dx} \text{sech } x,$ or

_____ $= - \text{sech } x \tanh x.$ Thus,

$\frac{dy}{dx} = -e^{-y}$ _____ $=$ _____ .

---

16. $\sinh u \frac{du}{dx}$ 17. $\cosh u \frac{du}{dx}$ 18. $\text{sech}^2 u \frac{du}{dx}$

19. $\sinh (3 - 2x^2),\ 3 - 2x^2,\ -12x \sinh^2 (3 - 2x^2) \cosh (3 - 2x^2)$

20. $\tanh 2x,\ \text{sech}^2 2x,\ 2x,\ e^x (2 \text{sech}^2 2x + \tanh 2x)$

21. $\sinh x \cdot \ln x,\ \sinh x,\ \frac{\sinh x}{x} + \cosh x \cdot \ln x,\ \sinh x + x \cosh x \ln x$

22. $e^y \frac{dy}{dx},\ \text{sech } x \tanh x,\ - \tanh x$

OBJECTIVE B : Integrate functions whose expressions involve hyperbolic functions.

23. $\int x \cosh (x^2 + 3) \, dx$

Let $u = x^2 + 3$, then $du =$ _____, and the integral becomes

$$\int x \cosh (x^2 + 3) \, dx = \int \underline{\hspace{3cm}} \, du = \underline{\hspace{3cm}} + C$$

$$= \underline{\hspace{4cm}}.$$

24. $\int \sinh^2 x \, dx$

From the identities $\cosh 2x = \sinh^2 x + \cosh^2 x$ and $\cosh^2 x - \sinh^2 x = 1$, we have

$\cosh 2x =$ _____ or $\sinh^2 x = \frac{1}{2} ($_____$)$.

Thus, $\int \sinh^2 x \, dx =$ _____ $+ C.$

25. $\int \tanh x \ln (\cosh x) \, dx$

Let $u = \ln (\cosh x)$, so $du =$ _____ and the integral becomes

$$\int \ln (\cosh x) \tanh x \, dx = \int \underline{\hspace{2cm}} \, du = \underline{\hspace{2cm}} + C$$

$$= \underline{\hspace{4cm}}.$$

9-3 HANGING CABLES.

26. To find an equation of a 50 foot long cable, weighing 5 lb/ft, and hanging under its own weight between two supports 30 feet apart, first choose the origin so that the y axis goes through the lowest point of the cable. Then an equation of the cable has the form

$y =$ _____ and $\frac{dy}{dx} =$ _____.

The length of the cable from $x = 0$ to $x = L$ is

$$s = \int_0^L \sqrt{1 + \left(\frac{dy}{dx}\right)^2} \, dx = \int_0^L \sqrt{\underline{\hspace{3cm}}} \, dx$$

$$= \int_0^L \underline{\hspace{3cm}} = \frac{H}{w} \sinh \left(\frac{w}{H}x\right)\Big]_0^L = \underline{\hspace{3cm}}.$$

Substituting the conditions $w = 5$, $L = 15$, $s = 25$ gives

---

23. $2x \, dx$, $\frac{1}{2} \cosh u$, $\frac{1}{2} \sinh u$, $\frac{1}{2} \sinh (x^2 + 3) + C$

24. $2 \sinh^2 x + 1$, $\cosh 2x - 1$, $\frac{1}{4} \sinh 2x - \frac{1}{2} x$

25. $\frac{1}{\cosh x} \cdot \sinh x \, dx$, u, $\frac{1}{2} u^2$, $\frac{1}{2} \ln^2 (\cosh x) + C$

25 = _____   or $\frac{125}{H} = \sinh \frac{75}{H}$.

Using a table of values of   sinh z,   the approximate value of
H   which will satisfy this last equation is   H = 40.8   lb.
Thus, an equation for the hanging cable is

$$y = \underline{\hspace{4cm}}.$$

27.   The lowest point on the cable in Problem 26 is the point
_____.   The highest point occurs when   x = 15   ft., so
y = _____,   or   y ≈ (8.16)(3.22) = 26.28   ft.
Therefore, the amount of sag in the cable is

$$sag = 26.28 - \underline{\hspace{2cm}} = \underline{\hspace{2.5cm}} \text{ feet.}$$

9-4   INVERSE HYPERBOLIC FUNCTIONS.

28. $y = \sinh^{-1} x$ means   _____.   Thus $x = \dfrac{e^y - \underline{\hspace{1cm}}}{2}$ or

$2xe^y = $ _____   or $e^{2y} - $ _____ $- 1 = 0$.   Solution of
this quadratic equation by the quadratic formula gives,

$e^y = \dfrac{2x \pm \sqrt{\underline{\hspace{1.5cm}}}}{2}$.   Since $e^y > 0$ we must have

$e^y = $ _____   or $\sinh^{-1} x = y = $ _____.

29.   We can use the formula found in Problem 28 to calculate $\dfrac{dy}{dx}$
for $y = \sinh^{-1} x$:

$$\frac{d}{dx} \ln \left(x + \sqrt{x^2 + 1} \right) = \frac{1}{x + \sqrt{x^2 + 1}} \cdot \frac{d}{dx} \left(\underline{\hspace{2cm}} \right)$$

$$= \frac{1}{x + \sqrt{x^2 + 1}} \cdot \left(\underline{\hspace{2cm}} \right)$$

$$= \frac{1}{x + \sqrt{x^2 + 1}} \cdot \left(\frac{\underline{\hspace{1.5cm}}}{\sqrt{x^2 + 1}} \right) = \underline{\hspace{1cm}}.$$

30.   An alternate way to calculate the derivative of $y = \sinh^{-1} x$
is as follows: Differentiate $x = \sinh y$ implicitly:

$\dfrac{d}{dx} x = \dfrac{d}{dx} \sinh y$ or $1 = $ _____.   Thus,

$$\frac{dy}{dx} = \frac{1}{\underline{\hspace{1cm}}} = \frac{1}{\sqrt{1 + \underline{\hspace{1cm}}}} = \underline{\hspace{1.5cm}}.$$

The positive square root is taken in the penultimate step
because   cosh y   is always   _____.

---

26. $\frac{H}{w} \cosh \left(\frac{w}{H} x \right)$, $\sinh \left(\frac{w}{H} x \right)$, $1 + \sinh^2 \left(\frac{w}{H} x \right)$, $\cosh \left(\frac{w}{H} x \right)$, $\frac{H}{w} \sinh \frac{wL}{H}$, $\frac{H}{5} \sinh \frac{5}{H}$ 15, $8.16 \cosh \frac{x}{8.16}$

27. $(0, 8.16)$, $8.16 \cosh \frac{15}{8.16}$, 8.16, 18.12

28. $x = \sinh y$, e^{-y}, $e^{2y} - 1$, $2xe^y$, $4x^2 + 4$, $x + \sqrt{x^2 + 1}$, $\ln \left(x + \sqrt{x^2 + 1} \right)$

29. $x + \sqrt{x^2 + 1}$, $1 + \dfrac{2x}{2\sqrt{x^2 + 1}}$, $\sqrt{x^2 + 1} + x$, $\dfrac{1}{\sqrt{x^2 + 1}}$

30. $\cosh y \cdot \dfrac{dy}{dx}$, $\cosh y$, $\sinh^2 y$, $\dfrac{1}{\sqrt{1 + x^2}}$,   positive

OBJECTIVE A : Calculate the derivatives of functions expressed in terms of inverse hyperbolic functions.

31. $\frac{d}{dx} \tanh^{-1} e^x = \underline{\phantom{\frac{1}{xx}}} \cdot \frac{d}{dx} e^x = \underline{}.$

32. $\frac{d}{dx} \ln (\sinh^{-1} x) = \frac{1}{\sinh^{-1} x} \cdot \frac{d}{dx} \underline{} = \underline{}.$

33. $\frac{d}{dx} \sqrt{\coth^{-1} x} = \frac{1}{2} (\coth^{-1} x)^{-1/2} \cdot \frac{d}{dx} \underline{}$

    $= \underline{}.$

34. $\frac{d}{dx} \cosh^{-1} \frac{3}{x^2} = \underline{} \cdot \frac{-6}{x^3} = \underline{}.$

OBJECTIVE B : Evaluate integrals using integration formulas for inverse hyperbolic functions.

35. $\int_{-3}^{-2} \frac{dx}{\sqrt{x^2 + 1}}$

    Since $\int \frac{dx}{\sqrt{x^2 + 1}} = \underline{} + C = \ln (\underline{}) + C,$

    $\int_{-3}^{-2} \frac{dx}{\sqrt{x^2 + 1}} = \underline{}]_{-3}^{-2}$

    $= \ln (\sqrt{5} - 2) - \underline{} \approx 0.375.$

36. $\int_{0.5}^{0.9} \frac{dx}{x\sqrt{1 - x^2}}$

    For $0 < x < 1,$ $\int \frac{dx}{x\sqrt{1 - x^2}} = \underline{} + C.$

    Now, $\text{sech}^{-1} x = \cosh^{-1} \underline{} = \ln \left(\underline{}\right) = \ln \left(\frac{\underline{}}{x}\right).$
    Thus,

    $\int_{0.5}^{0.9} \frac{dx}{x\sqrt{1 - x^2}} = \underline{}]_{0.5}^{0.9}$

    $= -\ln \left(\frac{1 + \sqrt{1 - .81}}{.9}\right) + \underline{}$

    $= -\ln \left(\frac{10 + \sqrt{19}}{9}\right) + \underline{}$

    $\approx 0.850.$

---

31. $1 - e^{2x},$ $\dfrac{e^x}{1 - e^{2x}}$ 32. $\sinh^{-1} x,$ $\dfrac{1}{\sinh^{-1} x \cdot \sqrt{1 + x^2}}$ 33. $\coth^{-1} x,$ $\dfrac{1}{2(1 - x^2)\sqrt{\coth^{-1} x}}$

34. $\dfrac{1}{\sqrt{\frac{9}{x^4} - 1}},$ $\dfrac{-6}{x\sqrt{9 - x^4}}$ 35. $\sinh^{-1} x,$ $x + \sqrt{x^2 + 1},$ $\ln \left(x + \sqrt{x^2 + 1}\right),$ $\ln (\sqrt{10} - 3)$

36. $-\text{sech}^{-1} x,$ $\frac{1}{x},$ $\frac{1}{x} + \sqrt{\frac{1}{x^2} - 1},$ $1 + \sqrt{1 - x^2},$ $-\ln \left(\frac{1 + \sqrt{1 - x^2}}{x}\right),$ $\ln \left(\frac{1 + \sqrt{1 - .25}}{.5}\right),$ $\ln (2 + \sqrt{3})$

37. $\displaystyle\int \frac{dx}{16 - x^2} = \frac{1}{16} \int$ _____

Let $u = \frac{x}{4}$, $du =$ _____ , and the integral becomes

$\displaystyle\int \frac{dx}{16 - x^2} = \int \frac{4\ du}{\underline{\hspace{1cm}}} = \frac{1}{4} \left(\underline{\hspace{2.5cm}} \right) + C$

$\displaystyle = \underline{\hspace{4cm}} + C.$

---

37. $\dfrac{dx}{1 - (x^2/16)}$, $\frac{1}{4}\,dx$, $16(1 - u^2)$, $\frac{1}{2} \ln \left| \dfrac{1 + u}{1 - u} \right|$, $\frac{1}{8} \ln \left| \dfrac{4 + x}{4 - x} \right|$

CHAPTER 9   SELF-TEST

1. Define the hyperbolic function $y = \operatorname{csch} x$, and sketch its graph.

2. Given that $\cosh x = 2$, $x < 0$, find the values of the remaining hyperbolic functions at x.

Find $\dfrac{dy}{dx}$ in each of the following:

3. $y = \tanh (\sin x)$

4. $y = \coth^{-1} (\ln x)$

5. $y = \sqrt{\cosh^{-1} x^2}$

6. $y = \ln (\sinh x^3)$

7. $y = x^{-1} \tanh^{-1} x^2$

8. $y = \sinh^{-1} (\tan x)$

Integrate each of the following:

9. $\displaystyle \int \frac{4\ dx}{\left(e^x - e^{-x}\right)^2}$

10. $\displaystyle \int \frac{\sinh (\ln x)\ dx}{x}$

11. $\displaystyle \int \sqrt{1 + \cosh x}\ dx$

12. $\displaystyle \int_3^7 \frac{dx}{\sqrt{x^2 - 1}}$

13. $\displaystyle \int_0^{1/2} \frac{\cosh x\ dx}{1 - \sinh^2 x}$

14. $\displaystyle \int_0^1 \frac{dx}{\sqrt{e^{2x} + 1}}$

15. Find the length of the catenary $y = 3 \cosh \dfrac{x}{3}$ from $x = 0$ to $x = 3$.

SOLUTIONS TO CHAPTER 9 SELF-TEST

1. $y = \operatorname{csch} x = \dfrac{1}{\sinh x}$,

   where $\sinh x = \frac{1}{2} (e^x - e^{-x})$.
   The graph is sketched at the right.

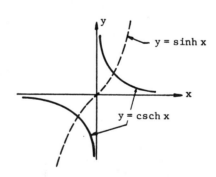

2. $\operatorname{sech} x = \dfrac{1}{\cosh x} = \frac{1}{2}$; $\tanh^2 x = 1 - \operatorname{sech}^2 x = \frac{3}{4}$ so that

   $\tanh x = \dfrac{-\sqrt{3}}{2}$. Since $x < 0$; $\sinh x = \cosh x \tanh x = -\sqrt{3}$;

   $\coth x = \dfrac{1}{\tanh x} = -\dfrac{2}{\sqrt{3}}$; and $\operatorname{csch} x = \dfrac{1}{\sinh x} = -\dfrac{1}{\sqrt{3}}$.

3. $\dfrac{dy}{dx} = \operatorname{sech}^2 (\sin x) \cdot \cos x$

4. $\dfrac{dy}{dx} = \dfrac{1}{[1 - (\ln x)^2]} \cdot \dfrac{1}{x}, \quad \ln x > 1$

5. $\dfrac{dy}{dx} = \dfrac{1}{2}\left(\cosh^{-1} x^2\right)^{-1/2} \cdot \dfrac{d}{dx}\left(\cosh^{-1} x^2\right)$

$\qquad = \dfrac{1}{2}\left(\cosh^{-1} x^2\right)^{-1/2} \cdot \dfrac{1}{\sqrt{x^4 - 1}} \cdot \dfrac{d}{dx}(x^2)$

$\qquad = \dfrac{x}{\sqrt{(x^4 - 1)\cosh^{-1} x^2}}, \quad x > 1$

6. $\dfrac{dy}{dx} = \dfrac{1}{\sinh x^3} \cdot \dfrac{d}{dx}(\sinh x^3) = \dfrac{1}{\sinh x^3} \cdot \cosh x^3 \cdot \dfrac{d}{dx}(x^3)$

$\qquad = 3x^2 \coth x^3$

7. $\dfrac{dy}{dx} = -\dfrac{1}{x^2} \tanh^{-1} x^2 + x^{-1} \cdot \dfrac{1}{1 - x^4} \cdot 2x$

$\qquad = -x^{-2} \tanh^{-1} x^2 + \dfrac{2}{1 - x^4}, \quad x^4 < 1$

8. $\dfrac{dy}{dx} = \dfrac{1}{\sqrt{\tan^2 x + 1}} \cdot \sec^2 x = \dfrac{\sec^2 x}{\sqrt{\sec^2 x}} = \sec x, \quad \text{if} \quad -\dfrac{\pi}{2} < x < \dfrac{\pi}{2}$

9. $\displaystyle\int \dfrac{4\,dx}{\left(e^x - e^{-x}\right)^2} = \int \operatorname{csch}^2 x\,dx = -\coth x + C$

10. Let $u = \ln x$, $du = \dfrac{1}{x}\,dx$, and the integral becomes

$\displaystyle\int \dfrac{\sinh(\ln x)\,dx}{x} = \int \sinh u\,du = \cosh u + C = \cosh(\ln x) + C$

11. From the identity $2\cosh^2 x = \cosh 2x + 1$, we find that $\sqrt{2}\cosh \dfrac{x}{2} = \sqrt{\cosh x + 1}$. Thus,

$\displaystyle\int \sqrt{\cosh x + 1}\,dx = \sqrt{2}\int \cosh \dfrac{x}{2}\,dx = 2\sqrt{2}\sinh \dfrac{x}{2} + C$

12. $\displaystyle\int_{-3}^{7} \dfrac{dx}{\sqrt{x^2 - 1}} = \cosh^{-1} x\Big]_3^7 = \ln\left(x + \sqrt{x^2 + 1}\right)\Big]_3^7$

$\qquad = \ln(7 + \sqrt{48}) - \ln(3 + \sqrt{8})$

$\qquad = \ln\left(\dfrac{7 + 4\sqrt{3}}{3 + 2\sqrt{2}}\right) \approx 2.39.$

13. Let $u = \sinh x$, $du = \cosh x\,dx$, so that

$\displaystyle\int_0^{1/2} \dfrac{\cosh x\,dx}{1 - \sinh^2 x} = \tanh^{-1}(\sinh x)\Big]_0^{1/2} = \dfrac{1}{2}\ln\left[\dfrac{1 + \sinh x}{1 - \sinh x}\right]_0^{1/2}$

$\qquad = \dfrac{1}{2}\ln\left[\dfrac{1 + 0.5211}{1 - 0.5211}\right] \approx 0.5778 \quad \text{by Tables.}$

14. $\displaystyle\int_0^1 \frac{dx}{\sqrt{e^{2x} + 1}} = \int_0^1 \frac{e^x \, dx}{e^x \sqrt{e^{2x} + 1}}$,   which is of the form

$$\int \frac{du}{|u| \sqrt{u^2 + 1}} = -\operatorname{csch}^{-1} u + C, \quad \text{for} \quad u = e^x \neq 0.$$

Thus, $\displaystyle\int_0^1 \frac{dx}{\sqrt{e^{2x} + 1}} = -\operatorname{csch}^{-1} e^x\Big]_0^1 = -\ln\left(\frac{1 + \sqrt{1 + e^{2x}}}{e^x}\right)\Big]_0^1$

$$= -\ln\left(\frac{1 + \sqrt{1 + e^2}}{e}\right) + \ln(1 + \sqrt{2}) \approx 0.52.$$

15. $\displaystyle s = \int_0^3 \sqrt{1 + \left(\frac{dy}{dx}\right)^2} \, dx = \int_0^3 \sqrt{1 + \sinh^2 \frac{x}{3}} \, dx$

$$= \int_0^3 \cosh \frac{x}{3} \, dx = 3 \sinh \frac{x}{3} \Big]_0^3 = 3 \sinh 1 \approx 3.53.$$

CHAPTER 10 POLAR COORDINATES

10-1 THE POLAR COORDINATE SYSTEM.

OBJECTIVE A : Given a point P in polar coordinates (r,θ), give the cartesian coordinates (x,y) of P.

1. The polar and cartesian coordinates are related by the equations x = _____ and y = _____ .

2. If P is the point $(-2,\frac{\pi}{6})$ in polar coordinates, then x = _____ and y = _____ so that P can be expressed in cartesian coordinates by (____ , ____) .

3. For $P = (-2,-\frac{\pi}{6})$ in polar coordinates, x = _____ and y = _____ so that P = (____ , ____) in cartesian coordinates.

OBJECTIVE B : Given a simple equation in polar coordinates, write an equivalent equation in cartesian coordinates and sketch the graph.

4. Consider the equation $r = -3 \sec \theta$. Then, $-3 = r$ _____ , or since x = _____ the equation is equivalent to _____ . This is an equation of a _____ line 3 units to the left of the _____ axis.

5. Consider the equation $r \sin (\theta - \frac{\pi}{3}) = \frac{1}{2}$. By the trigonometric summation identities, $\sin (\theta - \frac{\pi}{3}) = \sin \theta \cos \frac{\pi}{3} -$ _____

   = _____ .

   Therefore, the polar equation can be written

   $$\frac{1}{2} = \frac{1}{2} r \sin \theta - \text{_____} = \frac{1}{2} y - \text{_____} .$$

   Simplifying algebraically, y = _____ . This is an equation of a line with slope m = _____ and y-intercept b = _____ .

6. Suppose $r = \tan \theta \sec \theta$. Then, $r = \sin \theta \cdot$ _____ or $r = \dfrac{\sin \theta}{\text{\_\_\_\_}}$. Equivalently, $\sin \theta =$ _____ or

---

1. $r \cos \theta$, $r \sin \theta$ 2. $-2 \cos \frac{\pi}{6}$, $-2 \sin \frac{\pi}{6}$, $(-\sqrt{3}, -1)$ 3. $-2 \cos \left(-\frac{\pi}{6}\right)$, $-2 \sin \left(-\frac{\pi}{6}\right)$, $(-\sqrt{3}, 1)$

4. $\cos \theta$, $r \cos \theta$, $x = -3$, vertical, y

5. $\cos \theta \sin \frac{\pi}{3}$, $\frac{1}{2} \sin \theta - \frac{\sqrt{3}}{2} \cos \theta$, $\frac{\sqrt{3}}{2} r \cos \theta$, $\frac{\sqrt{3}}{2} x$, $\sqrt{3} x + 1$, $\sqrt{3}$, 1

r sin θ = _____.   In terms of cartesian coordinates the
equation becomes _____, which is readily recognized as an
equation of _____.

[OBJECTIVE C]:   Graph the points P(r,θ) whose polar coordinates
satisfy a given equation, inequality or inequalities.

7. $\theta = -\frac{\pi}{4}$, $-2 \leq r$
Sketch the graph at the right.

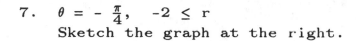

8. $r = 2$, $-\frac{3\pi}{4} < \theta \leq \frac{\pi}{6}$
Sketch the graph at the right.

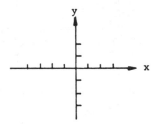

10-2   GRAPHING IN POLAR COORDINATES.

[OBJECTIVE]:   Given an equation F(r,θ) = 0 in polar coordinates,
analyze and sketch its graph.

9.   The graph of F(r,θ) = 0 is symmetric about the x-axis if the
equation is unchanged when _____ is replaced by _____.

10.   The graph of F(r,θ) = 0 is symmetric about the origin if the
equation is unchanged when _____ is replaced by _____.

11.   The graph of F(r,θ) = 0 is symmetric about the y-axis if the
equation is unchanged when _____ is replaced by _____.

---

6. $\sec^2 \theta$, $\cos^2 \theta$, $r \cos^2 \theta$, $r^2 \cos^2 \theta$, $y = x^2$, a parabola

7.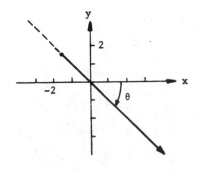

8.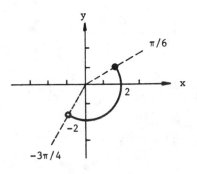

9. θ, $-\theta$ 10. r, $-$r   11. θ, $\pi - \theta$

12. Consider the curve given by
    r = 1 - 2 cos θ. Since
    cos (-θ) = cos θ, the curve
    is symmetric about the _____.

    Next, $\dfrac{dr}{d\theta}$ = _____. Thus,
    as θ varies from 0 to $\frac{\pi}{3}$,
    r increases from r = _____
    to r = _____; and as θ
    varies from $\frac{\pi}{2}$ to π, r
    increases from r = _____ to
    r = _____. Complete the
    following table of values for
    the curve, and sketch its graph
    using its symmetries.

θ	0	π/6	π/3	π/2	2π/3	5π/6	π
r							

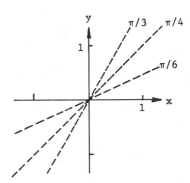

13. Consider the curve given by
    r^2 = sin θ. Since the sin θ
    must be nonnegative in order to
    equal the square of a real number,
    we must restrict θ to the
    interval _____. Since the
    equation remains unchanged when r
    is replaced by -r, the curve is
    symmetric about the _____.
    Also, since sin (π - θ) = sin θ,
    the curve is symmetric about the
    _____. Complete the following
    table and sketch the graph using
    these symmetries:

---

12. x-axis, 2 sin θ, -1, 0, 1, 3

θ	0	π/6	π/3	π/2	2π/3	5π/6	π
r	-1	1 - √3	0	1	2	1 + √3	3

The dashed portion of the curve is the rest
of it due to its symmetry about the x-axis.

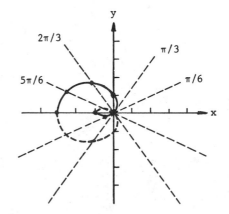

θ	0	$\pi/6$	$\pi/4$	$\pi/3$	$\pi/2$
r^2					
r					

10-3   POLAR EQUATIONS FOR CONICS AND OTHER CURVES.

OBJECTIVE A :   Given an equation in cartesian coordinates, write an equivalent equation in polar coordinates.

14.   Consider the equation of the ellipse $9x^2 + (y - 2)^2 = 4$. Expanding algebraically, and rearranging terms, we find $9x^2 + y^2 =$ _____ .   To transform the equation into polar coordinates, substitute   x = _____   and   y = _____   to obtain:

$$9r^2 \cos^2 \theta + \underline{\hspace{2cm}} = 4r \sin \theta, \quad \text{or}$$
$$9r^2(1 - \sin^2 \theta) + \underline{\hspace{2cm}} = 4r \sin \theta.$$

Simplifying algebraically,

$$r^2(\underline{\hspace{3cm}}) = 4r \sin \theta.$$

Either   r = 0,   or   r = _____ .   However, the latter equation includes the origin among its points and therefore represents the entire ellipse.

OBJECTIVE B :   Given an equation in polar coordinates, find an equivalent cartesian equation and sketch the graph.

15.   Given the equation $r = \dfrac{1}{3 \cos \theta + 2 \sin \theta}$,   clear fractions and obtain $3r \cos \theta +$ _____ = 1.   Next, substitute $x = r \cos \theta$ and   y = _____   to obtain   3x + _____ = 1. This is an equation of a straight line with slope   m = _____ and y-intercept   b = _____ .

---

13. $0 \le \theta \le \pi$, origin, y-axis

θ	0	$\pi/6$	$\pi/4$	$\pi/3$	$\pi/2$
r^2	0	1/2	$1/\sqrt{2}$	$\sqrt{3}/2$	1
r	0	±.71	±.84	±.93	±1

The portions of the graph in QII, QIII, and QIV are obtained by the symmetries.

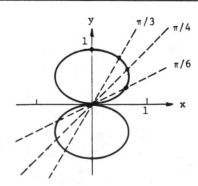

14.   4y, $r \cos \theta$, $r \sin \theta$, $r^2 \sin^2 \theta$, $r^2 \sin^2 \theta$, $9 - 8 \sin^2 \theta$, $\dfrac{4 \sin \theta}{9 - 8 \sin^2 \theta}$

15. $2r \sin \theta$, $r \sin \theta$, 2y, $-\dfrac{3}{2}$, $\dfrac{1}{2}$

16. The equation $r^2 - 5r + 4 = 0$ factors into $(r-4)(\underline{\hspace{1.5cm}}) = 0$. Thus, $r = 4$ or $r = \underline{\hspace{1cm}}$. The graph is two concentric circles, one of radius 4 and the other of radius 1, centered at the origin.

17. Consider $r^2 = 2 \csc 2\theta$. Then, $r^2 \sin 2\theta = \underline{\hspace{1.5cm}}$. Now, $\sin 2\theta = 2 \underline{\hspace{2cm}}$, so the equation becomes $2r \sin \theta \cdot \underline{\hspace{2cm}} = 2$ or $\underline{\hspace{1.5cm}} = 2$. That is, $xy = 1$ which is an equation of $\underline{\hspace{2.5cm}}$ with center $\underline{\hspace{2cm}}$.

18. Given the equation $r = \dfrac{3}{1 - 2 \cos \theta}$, clear fractions to obtain $r - \underline{\hspace{1.5cm}} = 3$; or substituting $x = r \cos \theta$, $r = \underline{\hspace{2cm}}$. Hence, $r^2 = \underline{\hspace{1.5cm}} = 9 + 12x + \underline{\hspace{1cm}}$. Since $r^2 = x^2 + y^2$ this last equation simplifies to $y^2 = 9 + 12x + \underline{\hspace{1.5cm}}$ or $y^2 = 3(x + 2)^2 + (\underline{\hspace{1cm}})$. Therefore, $(x + 2)^2 - \underline{\hspace{1.5cm}} = 1$. This is an equation of $\underline{\hspace{2cm}}$ with center $\underline{\hspace{2cm}}$.

19. For $r = 2 \sin \left(\theta + \frac{\pi}{4}\right)$ we can expand the right side by the summation formula for the sine: $r = 2 \sin \theta \cos \frac{\pi}{4} + \underline{\hspace{2.5cm}}$ or $r = \underline{\hspace{2.5cm}}$. Hence, $r^2 = \sqrt{2}\, r \sin \theta + \underline{\hspace{2.5cm}}$. Since $x^2 + y^2 = r^2$, $x = r \cos \theta$, and $y = r \sin \theta$, substitution and algebraic simplification yields

$\left(x - \frac{\sqrt{2}}{2}\right)^2 + \underline{\hspace{2.5cm}} = 1$. This equation represents

$\underline{\hspace{3cm}}$ with center $\underline{\hspace{2.5cm}}$ and $r = 1$.

10-4 INTEGRALS IN POLAR COORDINATES.

OBJECTIVE A : Find the total plane area enclosed by a polar graph $r = f(\theta)$ and the rays $\theta = \alpha$, $\theta = \beta$.

20. The area bounded by the polar curve $r = f(\theta)$ and the rays $\theta = \alpha$, $\theta = \beta$ is given by the integral

$$A = \underline{\hspace{4cm}}.$$

21. Find the area inside the larger loop and outside the smaller loop of the polar graph $r = 1 - 2 \cos \theta$ given in Problem 12. Solution. The graph of the curve is sketched in the figure below. That part of the curve traced out as θ varies from $\theta = 0$ to $\theta = \pi$ is drawn in with a broader ink stroke. Now,

---

16. $r - 1$, 1   17. 2, $\sin \theta \cos \theta$, $r \cos \theta$, $2xy$, a hyperbola, $C(0,0)$

18. $2r \cos \theta$, $3 + 2x$, $(3 + 2x)^2$, $4x^2$, $3x^2$, -3, $\dfrac{y^2}{3}$, a hyperbola, $C(-2,0)$

19. $2 \cos \theta \sin \frac{\pi}{4}$, $\sqrt{2} \sin \theta + \sqrt{2} \cos \theta$, $\sqrt{2}\, r \cos \theta$, $\left(y - \frac{\sqrt{2}}{2}\right)^2$, a circle, $C\left(\frac{\sqrt{2}}{2}, \frac{\sqrt{2}}{2}\right)$

20. $\int_{\alpha}^{\beta} \frac{1}{2}\left[f(\theta)\right]^2 d\theta$

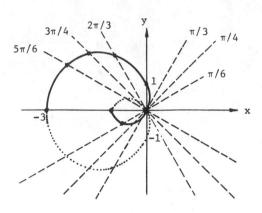

as θ varies from $\frac{\pi}{3}$ to π, the radius vector r sweeps out the larger loop of the curve including that portion of the smaller loop lying above the x-axis. By symmetry, the area of that smaller half-loop is the same as the area of the half-loop swept out by r as θ varies from 0 to $\frac{\pi}{3}$. Thus, the total area inside the larger loop and outside the smaller loop is

$$A = 2[\int_{\pi/3}^{\pi} \frac{1}{2} f^2(\theta) \ d\theta - \underline{\hspace{4cm}}]$$

$$= \int_{\pi/3}^{\pi} (1 - 4 \cos \theta + 4 \cos^2 \theta) \ d\theta - \underline{\hspace{5cm}}$$

$$= [\theta - 4 \sin \theta + 4(\frac{\theta}{2} + \frac{1}{4} \sin 2\theta]_{\pi/3}^{\pi} - \underline{\hspace{4cm}}$$

$$= (\pi + 2\pi) - [\frac{\pi}{3} - 4(\frac{\sqrt{3}}{2}) + 4 (\frac{\pi}{6} + \frac{1}{4} \cdot \frac{\sqrt{3}}{2})] - \underline{\hspace{3cm}}$$

$$= \underline{\hspace{3cm}} \approx 8.338.$$

OBJECTIVE B: Given a polar curve $r = f(\theta)$, calculate its arc length as θ varies from $\theta = a$ to $\theta = b$.

22. The differential element of arc length ds for the polar curve $r = f(\theta)$ satisfies the equation

$$ds^2 = \underline{\hspace{4cm}}.$$

Thus the length of arc traced out by the curve as θ varies from $\theta = a$ to $\theta = b$ is given by

$$s = \underline{\hspace{5cm}}.$$

23. To determine the length of the curve $r = 3 \sec \theta$ as θ varies from $\theta = 0$ to $\theta = \frac{\pi}{4}$, we find $\frac{dr}{d\theta} = \underline{\hspace{2cm}}$. Then the arc length is,

$$s = \int_0^{\pi/4} \sqrt{\underline{\hspace{3cm}}} \ d\theta = 3 \int_0^{\pi/4} \sec \theta \sqrt{\underline{\hspace{2cm}}} \ d\theta$$

$$= 3 \int_0^{\pi/4} \underline{\hspace{2cm}} \ d\theta = \underline{\hspace{2cm}}]_0^{\pi/4} = \underline{\hspace{2cm}}.$$

21. $\int_0^{\pi/3} \frac{1}{2} f^2(\theta) \ d\theta$, $\int_0^{\pi/3} (1 - 4 \cos \theta + 4 \cos^2 \theta) \ d\theta$, $[\theta - 4 \sin \theta + 4(\frac{\theta}{2} + \frac{1}{4} \sin 2\theta]_0^{\pi/3}$,

$[\frac{\pi}{3} - 4(\frac{\sqrt{3}}{2}) + 4(\frac{\pi}{6} + \frac{1}{4} \cdot \frac{\sqrt{3}}{2})]$, $3\sqrt{3} + \pi$

22. $r^2 \ d\theta^2 + dr^2$, $\int_a^b \sqrt{r^2 + (\frac{dr}{d\theta})^2} \ d\theta$

23. $3 \sec \theta \tan \theta$, $9 \sec^2 \theta + 9 \sec^2 \theta \tan^2 \theta$, $1 + \tan^2 \theta$, $\sec^2 \theta$, $3 \tan \theta$, 3

24. Consider the polar curve

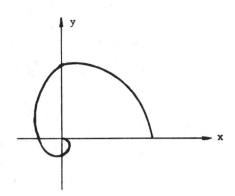

$r = \cos^4 \frac{\theta}{4}$. As θ varies
from $\theta = 0$ to $\theta = 2\pi$, the
equation describes the path
shown in the figure at the
right. As θ varies from
$\theta = 2\pi$ to $\theta = 4\pi$ the curve
shown is reflected across the
x-axis. Thus, the total arc
length is given by

$$s = 2 \int_0^{2\pi} \sqrt{r^2 + \underline{\hspace{2cm}}} \; d\theta .$$

Now, $\dfrac{dr}{d\theta} = \underline{\hspace{2cm}}$ so that

$$r^2 + \left(\frac{dr}{d\theta}\right)^2 = \cos^8 \frac{\theta}{4} + \underline{\hspace{2cm}} = \cos^6 \frac{\theta}{4}.$$

Thus, $s = 2 \int_0^{2\pi} \underline{\hspace{2cm}} \; d\theta .$

Since $\cos \frac{\theta}{4} \geq 0$ for $0 \leq \theta \leq 2\pi$,   the integral becomes

$$s = 2 \int_0^{2\pi} (1 - \sin^2 \frac{\theta}{4}) \; \underline{\hspace{2cm}} \; d\theta$$

$$= 2 \int \underline{\hspace{0.5cm}} \; \underline{\hspace{2cm}} \; du, \quad \text{where} \quad u = \sin \frac{\theta}{4}$$

$$= \underline{\hspace{1.5cm}} (\underline{\hspace{1.5cm}})] \underline{\hspace{0.5cm}} = \underline{\hspace{2cm}} .$$

OBJECTIVE C :   Find the area of the surface generated when a polar
graph is revolved about the x-axis or the y-axis.

25. The graph of the polar equation
$r = 5 \cos \theta$ is the circle shown
at the right. If the graph is
rotated about the x-axis, the
total surface area is generated
by that portion of the graph as
θ varies from $\theta = 0$ to
$\theta = \underline{\hspace{1.5cm}}$ because of the
symmetry of the graph across the x-axis.

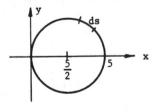

An element of arc length ds (see the figure) generates a
portion of surface area $dS = \underline{\hspace{2cm}}$ ds, where $y = r \sin \theta$
and $ds = \underline{\hspace{2cm}}$.   Thus,

$$dS = 2\pi \cdot \underline{\hspace{4cm}} \; d\theta = 2\pi (5 \cos \theta) \underline{\hspace{2cm}} \; d\theta .$$

Hence, the total surface area is given by

---

24. $\left(\dfrac{dr}{d\theta}\right)^2$, $- \cos^3 \frac{\theta}{4} \sin \frac{\theta}{4}$, $\cos^6 \frac{\theta}{4} \sin^2 \frac{\theta}{4}$, $\left|\cos^3 \frac{\theta}{4}\right|$, $\cos \frac{\theta}{4}$, $\int_0^1 4(1 - u^2) \; du$, $8(u - \frac{1}{3} u^3)]_0^1$, $\frac{16}{3}$

25. $\frac{\pi}{2}$, $2\pi y$, $\sqrt{dr^2 + r^2 \; d\theta^2}$, $r \sin \theta \sqrt{25 \sin^2 \theta + 25 \cos^2 \theta}$, $5 \sin \theta$,

$\int_0^{\pi/2} \sin \theta \cos \theta \; d\theta$, $50\pi \cdot \frac{1}{2} \sin^2 \theta]_0^{\pi/2}$, 25π

$$S = 50\pi \int_{\underline{\quad}}^{\overline{\quad}} \underline{\hspace{3cm}} \, d\theta = \underline{\hspace{4cm}}]\overline{\underline{\hspace{0.5cm}}} = \underline{\hspace{2cm}}$$

$$\approx 78.54.$$

26. If the graph in Problem 25 is rotated about the axis $\theta = \pi/2$, the total surface area is generated by the graph as θ varies from $\theta = 0$ to $\theta = \underline{\hspace{2cm}}$. An element of arc length ds now generates a portion of the surface area

$$dS = \underline{\hspace{3cm}} = 2\pi \left(\underline{\hspace{2cm}} \right) \cdot 5 \, d\theta.$$

Thus, the total surface area is given by

$$S = 50\pi \int_{\underline{\quad}}^{\overline{\quad}} \underline{\hspace{3cm}} \, d\theta = 50\pi \, [\underline{\hspace{3cm}}]\overline{\underline{\hspace{0.5cm}}}$$

$$= \underline{\hspace{3cm}} \approx 246.74.$$

---

26. π, $2\pi x \, ds$, $r \cos \theta$, $\displaystyle\int_0^\pi \cos^2 \theta \, d\theta$, $\frac{1}{2}\theta + \frac{1}{4}\sin 2\theta \,]_0^\pi$, $25\pi^2$

CHAPTER 10   SELF-TEST

1. Convert the following from polar coordinates to cartesian coordinates.

    (a) $\left(-6, \frac{\pi}{4}\right)$ (b) $\left(1, -\frac{5\pi}{6}\right)$ (c) $\left(-2, -\frac{7\pi}{12}\right)$

2. Write the following simple polar equations in cartesian form.

    (a) $r = 5$ (b) $\theta = \frac{3\pi}{4}$ (c) $r = -5 \csc \theta$

In Problems 3 and 4, graph the polar equation.

3. $r = 2 \cos 4\theta$ 4. $r^2 = -\sin 2\theta$

5. Determine a cartesian equation, and sketch the curve, for $r = \cos \theta + 5 \sin \theta$.

6. Find a polar equation of the line with slope $m = -2$ passing through the cartesian point $(1, -3)$.

7. Find a polar equation of the circle centered at the cartesian point $(\frac{1}{3}, 0)$ and passing through the origin.

8. Find the length of the polar curve $r = a \cos (\theta + b)$ from $\theta = 0$ to $\theta = \pi$, where a and b are constants.

9. Find the area of the region bounded on the outside by the graph of $r = 2 + 2 \sin \theta$ for $\theta = 0$ to $\theta = \pi$, and on the inside by the graph of $r = 2 \sin \theta$.

10. Write an integral expressing the surface area generated by rotating the portion of the polar curve $r = 1 + \cos \theta$ in the first quadrant about $\theta = \pi/2$.

SOLUTIONS TO CHAPTER 10 SELF-TEST

1. (a) $x = -6 \cos \frac{\pi}{4} = -6 \cdot \frac{\sqrt{2}}{2} = -3\sqrt{2}; \quad y = -6 \sin \frac{\pi}{4} = -3\sqrt{2}$

    (b) $x = 1 \cos \left(-\frac{5\pi}{6}\right) = \cos \frac{5\pi}{6} = -\frac{\sqrt{3}}{2};$

          $y = 1 \sin \left(-\frac{5\pi}{6}\right) = -\sin \frac{5\pi}{6} = -\frac{1}{2}$

    (c) $x = -2 \cos \left(-\frac{7\pi}{12}\right) = -2 \cos \frac{7\pi}{12} = \frac{\sqrt{2}}{2} (\sqrt{3} - 1) \approx 0.518$

          $y = -2 \sin \left(-\frac{7\pi}{12}\right) = 2 \sin \frac{7\pi}{12} = \frac{\sqrt{2}}{2} (\sqrt{3} + 1) \approx 1.932$

2. (a) $\pm \sqrt{x^2 + y^2} = 5$ or $x^2 + y^2 = 25$

    (b) $y = -x$

    (c) Equivalently, $r \sin \theta = -5$, or $y = -5$

3. $r = 2 \cos 4\theta$ is symmetric about
the x-axis, the y-axis, and the
origin.
The graph is the eight-leafed rose
sketched at the right.

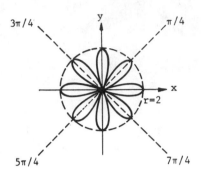

4. $r^2 = -\sin 2\theta$ is symmetric about
the origin since the equation is
unchanged when   r   is replaced by
-r. Notice that $-\sin 2\theta$ must
be nonnegative. If θ is
restricted to the interval $[0,2\pi]$,
then $-\sin 2\theta \geq 0$ if and only
if θ is in $[\frac{\pi}{2},\pi]$ or $[\frac{3\pi}{2},2\pi]$.
Using the following table and
symmetry we obtain the graph
sketched at the right:

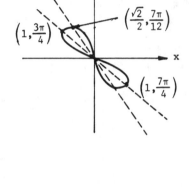

θ	$\pi/2$	$7\pi/12$	$3\pi/4$	π	$3\pi/2$
r	0	$\sqrt{2}/2$	1	0	0

The curve is a lemniscate.

5.   Since $r = 5 \cos \theta + 5 \sin \theta$,   for
$r \neq 0$ (r = 0   is on the graph at
$\theta = \frac{3\pi}{4}$), $r^2 = 5r \cos \theta + 5r \sin \theta$,

or $x^2 + y^2 = 5x + 5y$.   Then,
completing the squares in the   x
and   y   terms gives

$$\left(x - \frac{5}{2}\right)^2 + \left(y - \frac{5}{2}\right)^2 = \frac{25}{2}.$$

This is a circle with center
$\left(\frac{5}{2},\frac{5}{2}\right)$ and radius $\frac{5}{\sqrt{2}}$.

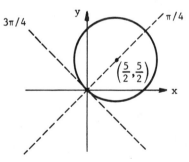

6.   A cartesian equation of the line is given by $y + 3 = -2(x - 1)$
or $2x + y = -1$.   Thus, $2r \cos \theta + r \sin \theta = -1$,   or solving
for   r, $r = \dfrac{-1}{2 \cos \theta + \sin \theta}$.

7. A cartesian equation of the circle is given by $\left(x - \frac{1}{3}\right)^2 + y^2 = \frac{1}{9}$ or $x^2 + y^2 = \frac{2}{3} x$. Thus, $3r^2 = 2r \cos \theta$, or since $r = 0$ lies on the graph $3r = 2 \cos \theta$ when $\theta = \frac{\pi}{2}$, the latter gives a polar equation of the circle.

8. For $r = a \cos (\theta + b)$, $\frac{dr}{d\theta} = -a \sin (\theta + b)$ so that $r^2 + \left(\frac{dr}{d\theta}\right)^2 = a^2 \cos^2 (\theta + b) + a^2 \sin^2 (\theta + b) = a^2$. Therefore the arc length is given by the integral $s = \int_0^{\pi} \sqrt{a^2} \, d\theta = |a|\pi$.

9. A graph depicting the region is shown in the figure at the right (the shaded portion represents the area we seek). Thus the area is given by

    $A = \frac{1}{2} \int_0^{\pi} (2 + 2 \sin \theta)^2 \, d\theta$

    $\quad - \frac{1}{2} \int_0^{\pi} (2 \sin \theta)^2 \, d\theta$

    $\quad = \frac{1}{2} \int_0^{\pi} (4 + 8 \sin \theta) \, d\theta$

    $\quad = 2(\theta - 2 \cos \theta)]_0^{\pi}$

    $\quad = 2\pi + 8 \approx 14.28.$

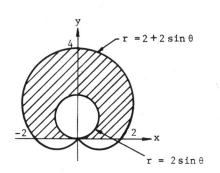

10. A sketch of the surface is shown in the figure at the right. An element of arc length ds generates a portion of surface area

         $dS = 2\pi x \, dx.$

    Now, $\frac{ds}{d\theta} = \sqrt{r^2 + \left(\frac{dr}{d\theta}\right)^2}$

    $\quad = \sqrt{1 + 2 \cos \theta + \cos^2 \theta) + \sin^2 \theta} = \sqrt{2} \sqrt{1 + \cos \theta}$

    Hence, $dS = 2\pi x \, ds = 2\pi r \cos \theta \, ds = 2\sqrt{2}\pi(1 + \cos \theta)^{3/2} \cos \theta \, d\theta$. Therefore, the total surface area generated is given by the integral

         $S = 2\sqrt{2}\pi \int_0^{\pi/2} (1 + \cos \theta)^{3/2} \cos \theta \, d\theta.$

    (The definite integral can be evaluated by using the identity $\cos^2 \frac{\theta}{2} = 1 + \cos \theta$, but it is tedious to carry out the calculations. Using Simpson's rule with $n = 12$, an approximate value to the integral is 39.31. This calculation was made using a calculator.)

NOTES.

CHAPTER 11 INFINITE SEQUENCES AND INFINITE SERIES

INTRODUCTION.

An ancient Greek paradox, due to the mathematician Zeno, concerned the following problem. Suppose that a man wants to walk a certain distance, say two miles, along a straight line from A to B. First he must pass the half-way point, then the 3/4 point, then the 7/8 point, and so on as illustrated in the following figure.

The fractional numbers in the figure indicate the distance in miles remaining to be covered. Therefore, on the assumption that a finite length contains an infinite number of points, the man must pass an infinite number of distance markers along the way. But that is impossible. The paradox is that the man does get to B, and in a finite amount of time, assuming he walks at some steady pace.

An analysis of the problem is not difficult. The total distance s from A to B is an infinite sum expressible as,

$$s = 1 + \left(\tfrac{1}{2}\right) + \left(\tfrac{1}{2}\right)^2 + \left(\tfrac{1}{2}\right)^3 + \cdots ,$$

and the paradox is dispelled if this infinite sum equals the finite number of 2 miles. That turns out to be exactly the case, as you will see further on in this chapter when it is established that

$$\frac{1}{1 - x} = 1 + x + x^2 + \cdots + x^n + \cdots , \quad \text{if} \quad |x| < 1.$$

For then,

$$2 = \frac{1}{1 - \tfrac{1}{2}} = 1 + \tfrac{1}{2} + \left(\tfrac{1}{2}\right)^2 + \left(\tfrac{1}{2}\right)^3 + \cdots + \left(\tfrac{1}{2}\right)^n + \cdots .$$

11-1 SEQUENCES OF NUMBERS.

OBJECTIVE A : Given a defining rule for the sequence $\{a_n\}$, write the first few items of the sequence.

1. A sequence is a _____ whose domain is the set of _____.

2. The numbers in the range of a sequence are called the _____ of the sequence. The number a_n is called the _____ of the sequence, or the term with _____ n.

---

1. function, positive integers   2. terms, nth term, index

3. For the sequence whose defining role is $a_n = 2 + \frac{1}{n}$, the first
    four terms are

    $a_1 = 3$, $a_2 = $ _____, $a_3 = $ _____,   and $a_4 = $ _____.

4. For the sequence whose defining rule is $a_n = \frac{4^n}{n!}$, the first
    four terms are

    $a_1 = 4$, $a_2 = $ _____, $a_3 = $ _____,   and $a_4 = $ _____.

5. For the sequence whose defining rule is $a_n = (-1)^{n+1}\left(\frac{n + 1}{n^3}\right)$,
    the first four terms are

    $a_1 = 2$, $a_2 = $ _____, $a_3 = $ _____,   and $a_4 = $ _____.

$\boxed{\text{OBJECTIVE B}}$:   Given a sequence $\{a_n\}$ determine if it converges or
                diverges.

6. The sequence $\{a_n\}$ converges to the number L if to every

    _____ ϵ there corresponds an _____ N such that

    _____.

    If no such limit L exists, we say that a_n _____.

7. If $0 < b < 1$, then $\{b^n\}$ converges to 0. To see why,
    consider the inequality
    $$|b^n - 0| = b^n < \epsilon.$$
    Thus, we seek an integer N such that

    _____ for all $n > N$.

    Since the natural logarithm $y = \ln x$ is an increasing
    function for all x,

    $b^n < \epsilon$ is equivalent to $n \ln b < \ln \epsilon$.

    Also, because $\ln b$ is a negative number for $0 < b < 1$, this
    latter inequality is equivalent to
    $$n > \text{_____}.$$
    Therefore, we need only choose an integer N satisfying

    _____,

    and the criterion set forth in Problem 6 for convergence to 0
    is satisfied.

11-2   LIMIT THEOREMS.

$\boxed{\text{OBJECTIVE}}$:   Given a sequence $\{a_n\}$ use the limit theorems to find
            the limit if the sequence converges.

---

3. $\frac{5}{2}$, $\frac{7}{3}$, $\frac{9}{4}$ 4. 8, $\frac{32}{3}$, $\frac{32}{3}$ 5. $-\frac{3}{8}$, $\frac{4}{27}$, $-\frac{5}{64}$

6. positive number, index, $|a_n - L| < \epsilon$ for all $n > N$, diverges   7. $b^n < \epsilon$, $\ln \epsilon / \ln b$, $N > \ln \epsilon / \ln b$

8. Consider the sequence defined by $a_n = \frac{4^n}{n!}$. Thus,

$$a_n = \frac{4 \cdot 4 \cdot 4 \cdot 4 \cdots 4}{1 \cdot 2 \cdot 3 \cdot 4 \cdots n} = \underline{\quad\quad} \left(\frac{4 \cdot 4 \cdots 4}{5 \cdot 6 \cdots n}\right) \quad \text{if} \quad n > 5$$

$$\leq \frac{32}{3} \left(\frac{4}{5}\right)\underline{\quad} = \left(\frac{32}{3}\right)\left(\frac{5}{4}\right)^4 \left(\frac{4}{5}\right)^n$$

Since $0 \leq a_n$ for all n, the Sandwich Theorem 3 on page 637 of the Thomas/Finney text gives $a_n \to \underline{\quad\quad}$ because $(4/5)^n \to 0$ from Problem 7.

9. For the sequence $\left\{\frac{n^3 + 5}{n^2 - 1}\right\}$, $\lim\limits_{n\to\infty} \frac{n^3 + 5}{n^2 - 1} = \lim\limits_{n\to\infty} \frac{n + (5/n^2)}{1 - (1/n^2)} = \underline{\quad\quad}$.

Therefore, the sequence $\underline{\quad\quad\quad\quad}$.

10. Let $a_n = \left(1 - \frac{1}{2^2}\right)\left(1 - \frac{1}{3^2}\right)\left(1 - \frac{1}{4^2}\right) \cdots \left(1 - \frac{1}{n^2}\right)$.

Thus, $\ln a_n = \sum\limits_{k=2}^{n} \ln\left(1 - \frac{1}{k^2}\right) = \sum\limits_{k=2}^{n} \ln\left(\frac{k^2 - 1}{k^2}\right)$

$$= \sum\limits_{k=2}^{n} [\ln(k + 1) + \ln(k - 1) - \underline{\quad\quad\quad}]$$

$$= -\ln 2 + [\ln 3 + \ln 1 - \ln 2]$$
$$+ [\ln 4 + \ln 2 - 2 \ln 3]$$
$$+ [\underline{\quad\quad\quad\quad\quad}]$$
$$+ [\ln 6 + \ln 4 - 2 \ln 5]$$
$$\vdots$$
$$+ [\ln n + \ln(n - 2) - 2 \ln(n - 1)]$$
$$+ [\ln(n + 1) + \ln(n - 1) - 2 \ln n]$$

Now, look closely at the expanded sum on the right. Notice that two terms contain $+\ln 3$ and one term contains $-2 \ln 3$, two terms contain $+\ln 4$ and one term contains $-2 \ln 4$, and so forth. Thus, most of the terms cancel each other out and we are left with,

$$\ln a_n = -\ln 2 + \underline{\quad\quad\quad\quad}.$$

Therefore,

$$\lim\limits_{n\to\infty} \ln a_n = \ln \frac{1}{2} + \lim\limits_{n\to\infty} \ln\left(\frac{n + 1}{n}\right) = \underline{\quad\quad\quad}.$$

By applying Theorem 4 on page 638 in the text to $f(x) = e^x$, we have

$$a_n = e^{\ln a_n} \to e^{\ln(1/2)} = \underline{\quad\quad} \quad \text{as} \quad n \to \infty.$$

11. For the sequence defined by $a_n = (-1)^{n+1}\left(\frac{n + 1}{n^3}\right)$,

$$0 \leq |a_n| = \underline{\quad\quad\quad} = \frac{1}{n^2} + \underline{\quad\quad\quad}.$$

---

8. $\frac{32}{3}$, n - 4, 0   9. ∞, diverges   10. $2 \ln k$, $\ln 5 + \ln 3 - 2 \ln 4$, $\ln(n + 1) - \ln n$, $\ln \frac{1}{2}$, $\frac{1}{2}$

Therefore,

$$(-1)^{n+1} \frac{n + 1}{n^3} \rightarrow \underline{\hspace{2cm}} \quad \text{as} \quad n \rightarrow \infty,$$

by the sequence version of the Sandwich Theorem.

11-3 LIMITS THAT ARISE FREQUENTLY.

OBJECTIVE : Given a sequence $\{a_n\}$, determine if it converges or diverges. If it converges, use the limits calculated in this article of the text, or logarithms or l'Hôpital's rule, to find its limit.

12. Let $a_n = \left(\frac{3n - 1}{5n + 1}\right)^n$. Then,

$$\frac{3n - 1}{5n + 1} < \frac{3n}{5n + 1} < \underline{\hspace{2cm}} \quad \text{implies} \quad 0 \le \left(\frac{3n - 1}{5n + 1}\right)^n < \underline{\hspace{2cm}}.$$

Therefore $a_n \rightarrow \underline{\hspace{1.5cm}}$ because $\left(\frac{3}{5}\right)^n \rightarrow 0$.

13. Let $a_n = \left(\frac{3n + 1}{5n - 1}\right)^{1/n}$. Then,

$\ln a_n = \frac{1}{n} (3n + 1) - \underline{\hspace{2.5cm}}.$
By l'Hôpital's rule,

$$\lim_{n \to \infty} \frac{\ln (3n + 1)}{n} = \lim_{n \to \infty} \frac{\underline{\hspace{1.5cm}}}{1} = \underline{\hspace{2cm}}, \quad \text{and}$$

$$\lim_{n \to \infty} \frac{\ln (5n - 1)}{n} = \lim_{n \to \infty} \frac{\underline{\hspace{1.5cm}}}{1} = \underline{\hspace{2cm}}.$$

Then, $\ln a_n \rightarrow 0$ so that $a_n = e^{\ln a_n} \rightarrow \underline{\hspace{2cm}}.$

14. Consider the sequence defined by $a_n = \left(1 + e^{-n}\right)^n$. Then,
$\ln a_n = \underline{\hspace{3cm}},$

and by l'Hôpital's rule,

$$\lim_{n \to \infty} \frac{\ln(1 + e^{-n})}{1/n} = \lim_{n \to \infty} \frac{\underline{\hspace{2cm}}}{-1/n^2} \quad (0/0)$$

$$= \lim_{n \to \infty} \frac{n^2}{e^n + 1} \quad (\infty/\infty)$$

$$= \lim_{n \to \infty} \underline{\hspace{2.5cm}} \quad (\text{still } \infty/\infty)$$

$$= \lim_{n \to \infty} \frac{2}{e^n} = \underline{\hspace{2cm}}.$$

Therefore,

$$a_n = e^{\ln a_n} \rightarrow \underline{\hspace{2cm}}.$$

---

11. $\frac{n + 1}{n^3}$, $\frac{1}{n^3}$, 0   12. $\frac{3}{5}$, $\left(\frac{3}{5}\right)^n$, 0

13. $\frac{1}{n} \ln (5n - 1)$, $3/(3n + 1)$, 0, $5/(5n - 1)$, 0, $e^0 = 1$

14. $n \ln (1 + e^{-n})$, $\left(\frac{1}{1 + e^{-n}}\right)\left(-e^{-n}\right)$, $\frac{2n}{e^n}$, 0, $e^0 = 1$

15. Let $a_n = \sqrt[n]{n^3}$. Then, $a_n = \left(n^3\right)^{1/n} = n\text{——} = \left(\sqrt[n]{n}\right)\text{——}$.
    Now, $\sqrt[n]{n} \rightarrow$ _____ by limit two in the text (see page 641),
    and if $f(x) = x^3$, then
    $$a_n = f\left(\sqrt[n]{n}\right) \rightarrow \text{——————}.$$

11-4 INFINITE SERIES.

16. If $\{a_n\}$ is a sequence, and
    $$s_n = a_1 + a_2 + \ldots + a_n,$$
    then the sequence $\{s_n\}$ is called an _____.

17. The number s_n is called the _____ of the series.

18. Instead of $\{s_n\}$ we usually use the notation _____ for the series.

19. The series $\sum\limits_{n=1}^{\infty} a_n$ is said to converge if the sequence

    _____ converges to a finite limit L. In that case we write

    _____ or $a_1 + a_2 + \ldots + a_n + \ldots = L$. If no such
    limit exists, the series is said to _____.

$\boxed{\text{OBJECTIVE A}}$: For a given geometric series $\sum\limits_{n=0}^{\infty} ar^n$, determine if

the series converges or diverges. If it does converge, then compute the sum of the series. The indexing of the series may be changed for a given problem.

20. Consider the series $\sum\limits_{n=0}^{\infty} \frac{3}{5^n}$. This is a geometric series with

    $a =$ _____ and $r =$ _____. Since $|r| < 1$, the geometric
    series _____, and its sum is given by
    $$\sum\limits_{n=0}^{\infty} \frac{3}{5^n} = \text{——————} = \text{——————}.$$

21. The series $\sum\limits_{n=2}^{\infty} (-1)^n \frac{4}{3^n}$ is a geometric series with $a =$ _____

    and $r =$ _____. Since $|r| < 1$, the geometric series
    _____. However, the index begins with $n = 2$ instead of
    $n = 0$. Now,

---

15. $\frac{3}{n}$, 3, 1, $1^3 = 1$ 16. infinite series   17. nth partial sum

18. $\sum\limits_{n=1}^{\infty} a_n$ 19. $\{s_n\}$, $\sum\limits_{n=1}^{\infty} a_n = L$, diverge

20. 3, $\frac{1}{5}$, converges, $\frac{3}{1 - \frac{1}{5}}$, $\frac{15}{4}$

$$\sum_{n=2}^{\infty} (-1)^n \frac{4}{3^n} = \sum_{n=0}^{\infty} (-1)^n \frac{4}{3^n} - (\underline{\hspace{2cm}})$$

$$= \underline{\hspace{3cm}} - \frac{8}{3} = \underline{\hspace{2cm}}.$$

22. The series $\sum_{n=3}^{\infty} \frac{2^n}{7}$ is a geometric series with $a = \underline{\hspace{2cm}}$ and $r = \underline{\hspace{1.5cm}}$. Since $\underline{\hspace{2cm}}$ the series diverges.

23. The repeating decimal $0.15\ 15\ 15\ \ldots$ is a geometric series in disguise. It can be written as

$$0.15\ 15\ 15\ \ldots = \frac{15}{100} + \frac{15}{\underline{\hspace{1cm}}} + \frac{15}{\underline{\hspace{1cm}}} + \ldots$$

$$= \sum_{n=1}^{\infty} \underline{\hspace{2cm}} = \sum_{n=0}^{\infty} \frac{15}{100^n} - \underline{\hspace{1.5cm}}$$

$$= \underline{\hspace{2cm}} - 15 = \frac{\underline{\hspace{2cm}}}{99}.$$

24. Sometimes the terms of a given series are a sum or difference of terms, each of which beongs to a geometric series. For example,

$$\sum_{n=0}^{\infty} \left(\frac{7}{3^n} - \frac{1}{2^n}\right) = \sum_{n=0}^{\infty} \frac{7}{3^n} - \sum_{n=0}^{\infty} \frac{1}{2^n}$$

$$= \underline{\hspace{2cm}} - \underline{\hspace{2cm}} = \underline{\hspace{2cm}} - \frac{4}{2} = \underline{\hspace{2cm}}.$$

[OBJECTIVE B]: Given an elementary series $\sum_{n=1}^{\infty} a_n$, determine whether it converges or diverges. If it converges, find the sum.

25. Consider the series $\sum_{n=1}^{\infty} \frac{1}{(2n-1)(2n+1)}$. This is not a geometric series. However, we can use partial fractions to re-write the kth term:

$$\frac{1}{(2k-1)(2k+1)} = \frac{1}{2}\left[\underline{\hspace{1.5cm}} - \frac{1}{2k+1}\right].$$

This permits us to write the partial sum

$$\sum_{n=1}^{k} \frac{1}{(2n-1)(2n+1)} = \frac{1}{1\cdot 3} + \frac{1}{3\cdot 5} + \ldots + \frac{1}{(2k-1)(2k+1)}$$

as

$$s_k = \frac{1}{2}\left(\frac{1}{1} - \frac{1}{3}\right) + \frac{1}{2}\left(\underline{\hspace{1.5cm}}\right) + \frac{1}{2}\left(\underline{\hspace{1.5cm}}\right)$$

$$+ \ldots + \frac{1}{2}\left(\frac{1}{2k-1} - \frac{1}{2k+1}\right).$$

21. 4, $-\frac{1}{3}$, converges, $4 - \frac{4}{3}$, $\frac{4}{1+\frac{1}{3}}$, $\frac{1}{3}$ 22. $\frac{1}{7}$, 2, $|r| > 1$

23. 100^2, 100^3, $\frac{15}{100^n}$, 15, $\frac{15}{1-\frac{1}{100}}$, $1500 - 1485$ or 15 24. $\frac{7}{1-\frac{1}{3}}$, $\frac{1}{1-\frac{1}{2}}$, $\frac{21}{2}$, $\frac{17}{2}$

By removing parentheses on the right, and combining terms, we find that

$$s_k = \text{_____} .$$

Therefore, $s_k \rightarrow$ _____ and the series _____ converge. Hence,

$$\sum_{n=1}^{\infty} \frac{1}{(2n-1)(2n+1)} = \text{_____} .$$

26. For the series $\sum_{n=1}^{\infty} \frac{n^n}{n!}$, we have for every index n,

$$a_n = \frac{n^n}{n!} = \frac{n \cdot n \cdot n \cdots n}{1 \cdot 2 \cdot 3 \cdots n} = \left(\frac{n}{1}\right) \left(\frac{n}{2}\right) \left(\text{_____}\right) \cdots \left(\frac{n}{n}\right) > \text{_____} .$$

Therefore, $\lim_{n \to \infty} a_n \neq 0$. We conclude that the series _____ .

27. Consider the series $\sum_{n=1}^{\infty} (-1)^{n+1} \left(1 + \frac{1}{3^n}\right)$. For large values of the index n, the absolute value of the nth term,

$$|a_n| = 1 + 3^{-n}$$

is close to _____ . Therefore, the limit

$$\lim_{n \to \infty} (-1)^{n+1} \left(1 + \frac{1}{3^n}\right) \text{_____ exist.}$$ We conclude that the series _____ .

28. For the series $\sum_{n=1}^{\infty} \left(\frac{n+2}{n}\right)^n$, we have

$$\lim_{n \to \infty} \left(\frac{n+2}{n}\right)^n = \lim_{n \to \infty} \left(1 + \frac{2}{n}\right)^n = \text{_____} .$$ Thus, the series _____ because the limit of the nth term is not zero.

29. The kth partial sum of the series $\sum_{n=1}^{\infty} [n^3 - (n+1)^3]$ is

$$s_k = (1 - 2^3) + \text{_____} .$$ By removing parentheses on the right, and combining terms, we find that $s_k = $ _____ . Therefore, the number s_k is less than or equal to $-k^3$ at each stage, so we conclude that the series _____ .

---

25. $\frac{1}{2k-1}$, $\frac{1}{3} - \frac{1}{5}$, $\frac{1}{5} - \frac{1}{7}$, $\frac{1}{2}\left(1 - \frac{1}{2k+1}\right)$, $\frac{1}{2}$, does, $\frac{1}{2}$ 26. $\frac{n}{3}$, 1, diverges

27. 1, does not, diverges   28. e^2, diverges

29. $(2^3 - 3^3) + \dots + \left[k^3 - (k+1)^3\right]$, $1 - (k+1)^3$, diverges

11-5   SERIES WITH NONNEGATIVE TERMS: COMPARISON AND INTEGRAL TESTS.

OBJECTIVE A :   Use one of the eight tests for divergence and
convergence of infinite series, listed on pages 667-
668 of the Thomas/Finney text, to determine whether a
given series converges or diverges.

30. $\sum\limits_{n=1}^{\infty} \dfrac{n + 5}{n^2 - 3n + 5}$

For every index n, $5n > -3n + 5$ because n is a positive
integer. Then, $n^2 + 5n > $ _____, and since $n^2 - 3n + 5$
is positive for every index it follows that
$$\dfrac{n + 2}{n^2 - 3n + 5} > \text{_____}.$$
We conclude that the given series _____ by comparison with
the series $\sum \dfrac{1}{n}$.

31. $\sum\limits_{n=2}^{\infty} \dfrac{1}{n(\ln n)^2}$

We apply the integral test $\int_2^{\infty} \dfrac{dx}{x(\ln x)^2} = \lim\limits_{b \to \infty} \underline{\hspace{2cm}} \Big]_b^2 = $ _____.

Therefore, the integral _____ and hence the series
_____ .

32. $\sum\limits_{n=1}^{\infty} \dfrac{(\ln n)^2}{n^3}$

The following argument shows that $\ln n < \sqrt{n}$ for every index
n. First, define the function $g(x) = \sqrt{x} - \ln x$. Now
$g'(x) = $ _____ is nonnegative if $x \geq$ _____, and it
follows that g is an increasing function of x for $x \geq 4$.
Also, $g(4) = 2 - \ln 4 \approx 0.613$ is positive. Therefore,
$g(x) > 0$ for $x \geq 4$. A simple verification using tables, or a
calculator, shows that $g(1)$, $g(2)$, and $g(3)$ are positive.
Hence, we have established that $\ln n < \sqrt{n}$ for every positive
integer n. Using this fact,
$$\dfrac{(\ln n)^2}{n^3} < \dfrac{(\sqrt{n})^2}{n^3} = \text{_____}.$$
We conclude that the series $\sum\limits_{n=1}^{\infty} \dfrac{(\ln n)^2}{n^3}$ _____ .

---

30. $n^2 - 3n + 5$, $\dfrac{1}{n}$, diverges   31. $\dfrac{1}{\ln x}$, $\dfrac{1}{\ln 2}$, converges, converges

32. $\dfrac{1}{2\sqrt{x}} - \dfrac{1}{x}$, 4, $\dfrac{1}{n^2}$, converges

OBJECTIVE B : Discuss the cardinal principle governing nondecreasing sequences and how it applies to infinite series of nonnegative terms.

33. A nondecreasing sequence $\{s_n\}$ is a sequence with the property that _____ for every n.

34. A sequence $\{s_n\}$ is said to be _____ from above if there is a finite constant M such that $s_n \leq M$ for every n.

35. If $\{s_n\}$ is a nondecreasing sequence that is bounded from above, then it _____.

36. If a nondecreasing sequence $\{s_n\}$ fails to be bounded, then it _____.

37. Applying the principle given in Problems 35 and 36 (see Theorem 10, page 659 of the text), if $\sum a_n$ is a series of nonnegative terms, then it converges if and only if the sequence of partial sums $\{s_n\}$ satisfies what property?

11-6   SERIES WITH NONNEGATIVE TERMS: RATIO AND ROOT TESTS

OBJECTIVE A : Given a series with nonnegative terms, determine convergence or divergence using the ratio and root tests.

38. $\sum\limits_{n=1}^{\infty} \dfrac{n!}{3^n}$

We try the ratio test. Thus, $\dfrac{a_{n+1}}{a_n} = \dfrac{(n+1)!/3^{n+1}}{n!/3^n} = $ _____.

Hence, $\rho = \lim\limits_{n \to \infty} \dfrac{a_{n+1}}{a_n} = $ _____, and the series _____.

39. $\sum\limits_{n=1}^{\infty} \left(\sqrt[n]{n} - 1 \right)^n$

We try the root test. Thus, for $a_n = \left(\sqrt[n]{n} - 1 \right)^n$,

$$\sqrt[n]{a_n} = \underline{\qquad} \to \underline{\qquad}.$$

Because $\rho = $ _____ we conclude that the series _____ according to the root test.

---

33. $s_n \leq s_{n+1}$ 34. bounded   35. converges

36. diverges to plus infinity   37. having an upper bound, as in Problem 34

38. $\frac{1}{3} \cdot (n+1)$, $+\infty$, diverges   39. $\sqrt[n]{n} - 1$, 0, 0, converges

OBJECTIVE B : Given a series whose nth term is an elementary expression containing some power of the variable x, find all values of x for which the series will converge. Begin with the ratio test or the root test, and then apply other tests as needed.

40. $\sum_{n=1}^{\infty} \frac{|x|^n}{\sqrt{n}3^n}$

The nth term of the series is $a_n = \frac{|x|^n}{\sqrt{n}3^n}$. If we apply the root test,

$$\sqrt[n]{a_n} = \frac{|x|}{\sqrt[n]{\sqrt{n}} \cdot 3} \rightarrow \underline{\hspace{2cm}}.$$

The root test therefore tells us that the series converges if $|x|$ is less than _____. We don't know what happens when $|x| = 3$. However, in that case the series becomes

$$\sum_{n=1}^{\infty} \frac{3^n}{\sqrt{n}3^n} = \sum_{n=1}^{\infty} \frac{1}{\sqrt{n}}.$$

This is a p-series with p = _____, and therefore the series _____. Thus, the original series

$\sum_{n=1}^{\infty} \frac{|x|^n}{\sqrt{n}3^n}$, converges for all values of x satisfying _____.

11-7 ABSOLUTE CONVERGENCE.

41. A series $\sum_{n=1}^{\infty} a_n$ is said to converge absolutely if

_____.

42. True or False:
(a) If a series converges absolutely, then it converges.
(b) If a series converges, then it converges absolutely.

OBJECTIVE : Given an infinite series containing (possibly) negative terms, determine whether it converges absolutely. Use the tests for convergence and divergence listed on page 676 of the text.

43. $\sum_{n=1}^{\infty} \frac{(-1)^{n+1} n \ln n}{3^n}$

The absolute value of the nth term of the series is $|a_n| = \frac{n \ln n}{3^n}$. Applying the ratio test,

$$\frac{|a_{n+1}|}{|a_n|} = \frac{(n+1) \ln (n+1)/3^{n+1}}{n \ln n/3^n} = \underline{\hspace{2cm}}. \text{ By l'Hôpital's}$$

rule, $\lim_{n \to \infty} \frac{\ln (n+1)}{\ln n} = \lim_{n \to \infty} \underline{\hspace{2cm}} = \underline{\hspace{1.5cm}}.$

---

40. $\frac{|x|}{3}$, 3, $\frac{1}{2}$, diverges, $|x| < 3$ 41. $\sum_{n=1}^{\infty} |a_n|$ converges   42. (a) True (b) False

Therefore, $\lim\limits_{n\to\infty} \dfrac{|a_{n+1}|}{|a_n|} = $ _____ , so we conclude that

$\sum\limits_{n=1}^{\infty} \dfrac{(-1)^{n+1}\, n \ln n}{3^n}$ _____ converge absolutely.

44. $\sum\limits_{n=1}^{\infty} \dfrac{2n - n^2}{n^3}$

The nth term of the series is $a_n = \dfrac{2n - n^2}{n^3} = \dfrac{n(2 - n)}{n^3}$ which is negative if $n >$ _____ . Thus, $-a_n = \dfrac{n - 2}{n^2}$ is positive for $n > 2$. Now, $\dfrac{n - 2}{n^2} > \dfrac{1}{2n}$ whenever $n >$ _____ . Since the series $\sum\limits_{n=1}^{\infty} \dfrac{1}{2n}$ _____ , we conclude that the series

$\sum\limits_{n=1}^{\infty} |a_n| = \sum\limits_{n=1}^{\infty} \dfrac{n - 2}{n^2}$ _____ by the comparison test.

Therefore, the original series _____ converge absolutely.

11-8   ALTERNATING SERIES AND CONDITIONAL CONVERGENCE.

OBJECTIVE A : Given an alternating series, determine if it converges or diverges. A test for convergence is Leibniz's Theorem on page 684 of the text, but another test may be required.

45. $\sum\limits_{n=1}^{\infty} (-1)^{n+1} \dfrac{n}{n^2 + 1}$

First we see that $a_n = \dfrac{n}{n^2 + 1}$ is positive for every n. Also, $\lim\limits_{n\to\infty} a_n = $ _____ . Next, we compare a_{n+1} with a_n for arbitrary n. Now, $\dfrac{n + 1}{(n + 1)^2 + 1} \le \dfrac{n}{n^2 + 1}$ if and only if

$(n + 1)(n^2 + 1) \le n[(n + 1)^2 + 1]$. This last inequality is equivalent to $n^3 + n^2 + n + 1 \le$ _____ or, $1 \le n^2 + n$ which is true. We conclude that the alternating series _____ converge by Leibniz's Theorem.

46. $\sum\limits_{n=1}^{\infty} (-1)^{n+1} \dfrac{1}{n^{1 + 1/n}}$

It is clear that $a_n = \dfrac{1}{n^{1 + 1/n}} = \dfrac{1}{n \cdot \sqrt[n]{n}}$ is positive for every n. Also, $\lim\limits_{n\to\infty} a_n = \lim\limits_{n\to\infty} \dfrac{1}{n} \cdot \lim\limits_{n\to\infty} \dfrac{1}{\sqrt[n]{n}} = $ _____ . We would

---

43. $\frac{1}{3}\left(\dfrac{n + 1}{n}\right) \dfrac{\ln(n + 1)}{\ln n}$, $\dfrac{1/(n + 1)}{1/n}$, 1, $\frac{1}{3}$, does

44. 2, 4, diverges, diverges, does not   45. 0, $n^3 + 2n^2 + 2n$, does

like to show that $\{a_n\}$ is a _____ sequence. One way is to replace n by the continuous variable x and show that the resultant function

$$y = f(x) = x^{1+ 1/x}, \quad x > 0,$$

which is the reciprocal of a_n for $x = n$, is an <u>increasing</u> function of x for every x. If we take the <u>logarithm of both</u> sides of this last equation, and differentiate implicitly with respect to x, we obtain

$$\ln y = \left(1 + \tfrac{1}{x}\right) \ln x, \quad \text{and} \quad \frac{y'}{y} = \frac{1}{x^2}\ (\underline{}) + \tfrac{1}{x}.$$

Simplifying algebraically, $y' = \frac{y}{x^2}(1 + \underline{} - \ln x)$.

Thus, since $x > \ln x$ for all $x > 0$, we find that y' is positive, and $y = f(x)$ is increasing for all x (so the reciprocal is _____). Therefore we have established that

$$\frac{1}{(n + 1) \cdot \sqrt[n+1]{n + 1}} < \frac{1}{n \cdot \sqrt[n]{n}} \quad \text{for every index } n.$$

We conclude that the alternating series _____.

OBJECTIVE B : Given an infinite series, use the tests studied in this chapter of the text to determine if the series is absolutely convergent, conditionally convergent, or divergent.

47. If a series $\sum a_n$ converges, but the series of absolute values $\sum |a_n|$ diverges, we say that the original series $\sum a_n$ is

_____.

48. In Problem 46 above, the series $\sum\limits_{n=1}^{\infty} (-1)^{n+1}\ \dfrac{1}{n^{1+ 1/n}}$ was shown to be convergent. We want to know if the series converges absolutely. Using the same technique as in Problem 46, it is easy to establish that

$$\sqrt[n+1]{n + 1} < \sqrt[n]{n} \quad \text{if} \quad n \geq 3:$$

we define the function $y = x^{1/x}$, $x \geq 3$, and show that y' is always negative; whence we conclude that y is a _____ function of x. In particular, $\sqrt[n]{n} < \sqrt[3]{3} \approx 1.44 < \tfrac{3}{2}$ if $n \geq$ _____. It follows that

$$\frac{1}{n \cdot \sqrt[n]{n}} > \underline{} \quad \text{if} \quad n \geq 4.$$

Since the harmonic series $\sum \tfrac{1}{n}$ diverges, we find that the series $\sum\limits_{n=1}^{\infty} \dfrac{1}{n^{1+ 1/n}}$ _____ by the comparison test.

---

46. $0 \cdot 1 = 0$, decreasing, $1 - \ln x$, x, decreasing, converges   47. conditionally convergent

Therefore, the original series $\sum_{n=1}^{\infty} (-1)^{n+1} \frac{1}{n^{1+1/n}}$ is

_____.

49. $\sum_{n=1}^{\infty} (-1)^{n+1} \left(\frac{n+1}{n}\right)^n$

In this case, $a_n = \left(\frac{n+1}{n}\right)^n = \left(1 + \frac{1}{n}\right)^n \rightarrow$ _____. Therefore

the series _____.

50. $\sum_{n=2}^{\infty} \frac{\cos n\pi}{n\sqrt{\ln n}}$

Since $\cos n\pi$ is 1 when n is even, and -1 when n is odd, the series is alternating in sign. Let us see if the series converges absolutely. Now,

$|a_n| = \left|\frac{\cos n\pi}{n\sqrt{\ln n}}\right| =$ _____. Applying the integral test,

$\int_1^{\infty} \frac{dx}{x\sqrt{\ln x}} = \lim_{b \to \infty}$ _____$\Big]_1^b =$ _____.

Therefore, the series $\sum |a_n|$ _____. To see if the original series converges we check the three conditions of Leibniz's Theorem (remember that the numerator $\cos n\pi$ simply determines the <u>sign</u> of the nth term of the series):

$\frac{1}{n\sqrt{\ln n}}$ is positive for all n, and converges to _____.

It is clear that $\frac{1}{(n+1)\sqrt{\ln(n+1)}} < \frac{1}{n\sqrt{\ln n}}$ because

$y = \ln x$ is an _____ function of x. Therefore, the

series $\sum_{n=1}^{\infty} \frac{\cos n\pi}{n\sqrt{\ln n}}$ ia _____.

|OBJECTIVE C|: Use the Alternating Series Estimation Theorem to estimate the magnitude of the error if the first k terms, for some specified number k, are used to approximate a given alternating series.

51.   It can be shown, with a little work, that the alternating harmonic series

$$\sum_{n=1}^{\infty} \frac{(-1)^{n+1}}{n}$$

converges to ln 2. If we wish to approximate ln 2 correct to four decimal places using this series, the alternating series error estimation gives

---

48. decreasing, 4, $\frac{2}{3n}$, diverges, conditionally convergent   49. e, diverges

50. $\frac{1}{n\sqrt{\ln n}}$, $2\sqrt{\ln x}$, $+\infty$, diverges, 0, increasing, conditinoally convergent

_____ $< 0.5 \times 10^{-5}$,   or $n >$ _____.

Therefore, we would need to sum the first 200,000 terms of the alternating harmonic series to <u>ensure</u> four decimal place accuracy in approximating $\ln 2$. This does not mean that fewer terms would <u>not</u> provide that accuracy. A more efficient approximation for $\ln 2$, accurate to four decimal places, uses Simpson's rule with $n = 6$ to estimate

$$\int_1^2 \frac{dx}{x}.$$

11-9   RECAPITULATION.

$\boxed{\text{OBJECTIVE}}$:   Use the convergence and divergence tests to determine if a given series converges.

52. $\sum\limits_{n=1}^{\infty} a_n$ with $a_1 = 2$ and $a_{n+1} = \dfrac{a_n}{n}$.   Using the recursion formula,

$a_{n+1} = \dfrac{a_n}{n} =$ _____ $a_{n-1} =$ _____ a_2.
Thus,

$$a_{n+1} = \frac{1}{n!}\, a_1 = \text{_____}.$$

The infinite series is $\sum\limits_{n=1}^{\infty} a_n = \sum\limits_{n=0}^{\infty} a_{n+1} = \sum\limits_{n=0}^{\infty} \dfrac{2}{n!}$.   The series

_____ by the _____ test.

11-10   ESTIMATING THE SUM OF A SERIES OF CONSTANT TERMS.

53.   The two possible sources of error in estimating the sum of a series are _____ errors and _____ errors.

54.   Truncation error is the difference between _____ _____.

55.   Round-off error is due to the fact that arithmetic on computers is done only on finitely many _____.

---

51. $\frac{1}{n}$, 2×10^5

52. $\frac{1}{n} \cdot \frac{1}{n-1}$, $\frac{1}{n} \cdot \frac{1}{n-1} \cdots \frac{1}{2}$, $\frac{2}{n!}$, converges, ratio   53.   truncation, round-off

54.   the sum of the series and its nth partial sum   55.   decimal or binary places

CHAPTER 11   SELF-TEST

1. Determine if each sequence $\{a_n\}$ converges or diverges. Find the limit of the sequence if it does converge.

   (a) $a_n = \sqrt{n + 1} - \sqrt{n}$

   (b) $a_n = \dfrac{1 + (-1)^n}{n\sqrt{n}}$

   (c) $a_n = \left(\dfrac{n - 0.05}{n}\right)^n$

   (d) $a_n = \dfrac{2^n}{5^{3 + 1/n}}$

2. Find the sum of each series.

   (a) $\displaystyle\sum_{n=0}^{\infty} (-1)^n \dfrac{3}{5^n}$

   (b) $\displaystyle\sum_{n=4}^{\infty} \dfrac{2}{(4n - 3)(4n + 1)}$

   (c) $\displaystyle\sum_{n=0}^{\infty} \left(\dfrac{5}{3^n} - \dfrac{2}{7^n}\right)$

   (d) $\dfrac{127}{1000} + \dfrac{127}{1000^2} + \dfrac{127}{1000^3} + \cdots + \dfrac{127}{1000^n} + \cdots$

In Problems 3-8, determine whether the given series converges or diverges. In each case, give a reason for your answer.

3. $\displaystyle\sum_{n=1}^{\infty} \dfrac{\sqrt{n}}{n^2 + 3}$

4. $\displaystyle\sum_{n=1}^{\infty} \dfrac{n!\,3^n}{10^n}$

5. $\displaystyle\sum_{n=1}^{\infty} \sin\left(\dfrac{n\pi - 2}{3n}\right)$

6. $\displaystyle\sum_{n=1}^{\infty} \left(\dfrac{n}{2n + 5}\right)^n$

7. $\displaystyle\sum_{n=1}^{\infty} \dfrac{1}{n + \sqrt{n}}$

8. $\displaystyle\sum_{n=1}^{\infty} \dfrac{\tan^{-1} n}{n^2 + 1}$

9. Find all values of x for which the given series converge.

   (a) $\displaystyle\sum_{n=1}^{\infty} (2x - 1)^{n!}$

   (b) $\displaystyle\sum_{n=2}^{\infty} \dfrac{\ln n}{n} x^n$

In Problems 10-13, determine whether the series are absolutely convergent, conditionally convergent, or divergent.

10. $\displaystyle\sum_{n=1}^{\infty} (-1)^{n+1} \dfrac{\sin n}{n^2 + 1}$

11. $\displaystyle\sum_{n=1}^{\infty} (-1)^{n+1} \dfrac{1}{(n + 1)^{1/n}}$

12. $\displaystyle\sum_{n=2}^{\infty} (-1)^n \dfrac{1}{(\ln n)^2}$

13. $\displaystyle\sum_{n=1}^{\infty} (-1)^{n+1} \dfrac{n + 1}{7n - 2}$

14. Estimate the magnitude of the error if the first five terms are used to approximate the series,

$$\sum_{n=1}^{\infty} (-1)^{n+1} \dfrac{2^n}{3^n}.$$

   Sum the first five terms, and state whether your approximation underestimates or overestimates the sum of the series.

SOLUTIONS TO CHAPTER 11 SELF-TEST

1. (a) $a_n = \sqrt{n+1} - \sqrt{n} = \dfrac{(\sqrt{n+1} - \sqrt{n})(\sqrt{n+1} + \sqrt{n})}{(\sqrt{n+1} + \sqrt{n})} = \dfrac{(n+1) - n}{\sqrt{n+1} + \sqrt{n}}$

$= \dfrac{1}{\sqrt{n+1} + \sqrt{n}} \rightarrow 0 \quad \text{as} \quad n \rightarrow \infty$

(b) $\sqrt[n]{n} \rightarrow 1$, but $1 + (-1)^n$ alternates back and forth between 0 and 2. Thus, for n large, a_n alternates between numbers very close to 2 and 0; hence the sequence diverges.

(c) $a_n = \left(\dfrac{n - 0.05}{n}\right)^n = \left(1 + \dfrac{-0.05}{n}\right)^n \rightarrow e^{-0.05} \approx 0.951$.

(d) $5^{3 + 1/n} = 125 \sqrt[n]{5} \rightarrow 125$, but $2^n \rightarrow +\infty$. Therefore, the sequence $\{a_n\}$ is unbounded and diverges.

2. (a) $\displaystyle\sum_{n=0}^{\infty} (-1)^n \dfrac{3}{5^n} = \sum_{n=0}^{\infty} 3 \left(-\dfrac{1}{5}\right)^n = \dfrac{3}{1 + \frac{1}{5}} = \dfrac{5}{2}$.

(b) Using the partial fraction decomposition,
$\dfrac{2}{(4k - 3)(4k + 1)} = \dfrac{1}{2}\left(\dfrac{1}{4k - 3}\right) - \dfrac{1}{2}\left(\dfrac{1}{4k + 1}\right)$, we write the partial sum

$$s_k = \sum_{n=4}^{\infty} \dfrac{2}{(4n - 3)(4n + 1)}$$

as
$$s_k = \dfrac{1}{2}\left(\dfrac{1}{13} - \dfrac{1}{17}\right) + \dfrac{1}{2}\left(\dfrac{1}{17} - \dfrac{1}{21}\right) + \dfrac{1}{2}\left(\dfrac{1}{21} - \dfrac{1}{25}\right) + \cdots$$
$$+ \dfrac{1}{2}\left(\dfrac{1}{4k - 3} - \dfrac{1}{4k + 1}\right).$$

Thus,
$$s_k = \dfrac{1}{2}\left(\dfrac{1}{13} - \dfrac{1}{4k + 1}\right) \rightarrow \dfrac{1}{26} \quad \text{as} \quad k \rightarrow \infty$$

so that
$$\sum_{n=4}^{\infty} \dfrac{2}{(4n - 3)(4n + 1)} = \dfrac{1}{26}.$$

(c) $\displaystyle\sum_{n=0}^{\infty} \left(\dfrac{5}{3^n} - \dfrac{2}{7^n}\right) = \sum_{n=0}^{\infty} \dfrac{5}{3^n} - \sum_{n=0}^{\infty} \dfrac{2}{7^n} = \dfrac{5}{1 - \frac{1}{3}} - \dfrac{2}{1 - \frac{1}{7}} = \dfrac{31}{6}$.

(d) $\displaystyle\sum_{n=1}^{\infty} 127 \left(\dfrac{1}{1000}\right)^n = \sum_{n=0}^{\infty} 127 \left(\dfrac{1}{1000}\right)^n - 127 = \dfrac{127}{1 - \frac{1}{1000}} - 127$

$= \dfrac{127,000 - 126,873}{999} = \dfrac{127}{999}$.

3. $\dfrac{\sqrt{n}}{n^2 + 3} < \dfrac{\sqrt{n}}{n^2} = \dfrac{1}{n^{3/2}}$ so that $\displaystyle\sum_{n=1}^{\infty} \dfrac{\sqrt{n}}{n^2 + 3}$ converges by comparison with the convergent p-series for $p = \dfrac{3}{2}$.

4. Using the ratio test, $\displaystyle\lim_{n\to\infty} \frac{(n+1)!3^{n+1}}{10^{n+1}} \cdot \frac{10^n}{n!3^n} = \lim_{n\to\infty} \frac{(n+1)3}{10}$ $= \infty$. Thus, $\displaystyle\sum_{n=1}^{\infty} \frac{n!3^n}{10^n}$ __diverges__ by the ratio test.

5. $\displaystyle\lim_{n\to\infty} \sin\left(\frac{n\pi - 2}{3n}\right) = \lim_{n\to\infty} \sin\left(\frac{\pi}{3} - \frac{2}{3n}\right) = \sin\frac{\pi}{3} = \frac{\sqrt{3}}{2} \neq 0$, so the series $\displaystyle\sum_{n=1}^{\infty} \sin\left(\frac{n\pi - 2}{3n}\right)$ __diverges__ by the nth-term test for divergence.

6. If $a_n = \left(\dfrac{n}{2n+5}\right)^n$, then $\sqrt[n]{a_n} = \dfrac{n}{2n+5} \to \dfrac{1}{2}$. Thus, the series $\displaystyle\sum_{n=1}^{\infty} \left(\frac{n}{2n+5}\right)^n$ __converges__ by the root test.

7. $\dfrac{1}{n+\sqrt{n}} > \dfrac{1}{n+n} = \dfrac{1}{2n}$ so that $\displaystyle\sum_{n=1}^{\infty} \frac{1}{n+\sqrt{n}}$ __diverges__ by comparison to the divergent series $\displaystyle\sum_{n=1}^{\infty} \frac{1}{2n}$.

8. $\displaystyle\int_1^{\infty} \frac{\tan^{-1} x \, dx}{x^2 + 1} = \lim_{b\to\infty} \frac{1}{2}\left(\tan^{-1} x\right)^2\Big]_1^b = \lim_{b\to\infty} \frac{1}{2}\left(\tan^{-1} b\right)^2 - \frac{1}{2}\tan^{-1} 1$
$= \frac{1}{2}\left(\frac{\pi}{2}\right)^2 - \frac{1}{2}\left(\frac{\pi}{4}\right)$

Therefore, the improper integral converges, so the original series $\displaystyle\sum_{n=1}^{\infty} \frac{\tan^{-1} n}{n^2 + 1}$ __converges__ by the integral test.

9. (a) If $|2x - 1| < 1$, then $|2x - 1|^{n!} < |2x - 1|^n$, for $n \geq 1$.
If $|2x - 1| \geq 1$, then $|2x - 1|^{n!} \geq |2x - 1|^n$, for $n \geq 1$.

Therefore, the series $\displaystyle\sum_{n=1}^{\infty} (2x - 1)^{n!}$ __converges__ absolutely for all values of x satisfying $|2x - 1| < 1$, or $0 < x < 1$ by comparison with the convergent geometric series $\displaystyle\sum_{n=1}^{\infty} (2x - 1)^n$. The series $\displaystyle\sum_{n=1}^{\infty} (2x - 1)^{n!}$ __diverges__ for all values of x satisfying $|2x - 1| \geq 1$ since $(2x - 1)^{n!} = |2x - 1|^{n!} \geq |2x - 1|^n$ if $n \geq 2$, and the geometric series $\displaystyle\sum_{n=1}^{\infty} (2x - 1)^n$ diverges for $|2x - 1| \geq 1$.

(b) Using the ratio test,
$\displaystyle\lim_{n\to\infty} \frac{\ln(n+1) \, |x|^{n+1}}{(n+1)} \cdot \frac{n}{\ln n \, |x|^n}$
$\displaystyle = \lim_{n\to\infty} \frac{\ln(n+1)}{\ln n} \cdot \frac{n+1}{n} \, |x| = |x|.$

Thus, the given power series converges absolutely for $|x| < 1$ and diverges for $|x| > 1$. We test the end-points of the interval.

For $x = 1$, the power series is $\sum\limits_{n=2}^{\infty} \frac{\ln n}{n}$. Now

$$\int_2^{\infty} \frac{\ln x}{x}\, dx = \lim_{b \to \infty} \frac{1}{2} (\ln x)^2 \big]_2^b = +\infty \quad \text{diverges, so the series}$$

$\sum\limits_{n=2}^{\infty} \frac{\ln n}{n}$ is divergent by the integral test.

For $x = -1$, the power series is $\sum\limits_{n=2}^{\infty} \frac{(-1)^n \ln n}{n}$. Since

$0 \le \lim\limits_{n \to \infty} \frac{\ln n}{n} \le \lim\limits_{n \to \infty} \frac{\sqrt{n}}{n} = 0$, and

$\frac{d}{dx}\left(\frac{\ln x}{x}\right) = \frac{1 - \ln x}{x^2} < 0$ for $x \ge 3$ implies that

$\frac{\ln (n + 1)}{n + 1} < \frac{\ln n}{n}$, the alternating series $\sum\limits_{n=2}^{\infty} \frac{(-1)^n \ln n}{n}$

converges by Leibniz's Theorem. Therefore, the power series

$\sum\limits_{n=2}^{\infty} \frac{\ln n}{n} x^n$ converges for all x satisfying $-1 \le x < 1$.

10. $\left|(-1)^{n+1} \frac{\sin n}{n^2 + 1}\right| \le \frac{1}{n^2 + 1}$, so the original series

$\sum\limits_{n=1}^{\infty} (-1)^{n+1} \frac{\sin n}{n^2 + 1}$ $\underline{\text{converges}}$ $\underline{\text{absolutely}}$ by the comparison test.

11. Since $\frac{1}{n^2} < \frac{1}{n + 1} < \frac{1}{n}$, it follows that

$\left(\frac{1}{\sqrt[n]{n}}\right)\left(\frac{1}{\sqrt[n]{n}}\right) < \frac{1}{\sqrt[n]{n + 1}} < \frac{1}{\sqrt[n]{n}}$. Thus, $\lim\limits_{n \to \infty} \frac{1}{\sqrt[n]{n + 1}} = 1$ so

the original series $\sum\limits_{n=1}^{\infty} (-1)^{n+1} \frac{1}{(n + 1)^{1/n}}$ $\underline{\text{diverges}}$ by the nth-term test.

12. $\lim\limits_{n \to \infty} \frac{1}{(\ln n)^2} = 0$, and $\frac{1}{[\ln (n + 1)]^2} < \frac{1}{(\ln n)^2}$ because $y = \ln x$ is an increasing function of x. Therefore the alternating

series $\sum\limits_{n=2}^{\infty} (-1)^n \frac{1}{(\ln n)^2}$ converges by Leibniz's Theorem.

However, since $\ln n < \sqrt{n}$ implies $\frac{1}{(\ln n)^2} > \frac{1}{n}$ if $n \ge 2$, the

series of absolute values $\sum\limits_{n=2}^{\infty} \frac{1}{(\ln n)^2}$ diverges by comparison with

the divergent harmonic series. Therefore, $\sum\limits_{n=2}^{\infty} (-1)^n \frac{1}{(\ln n)^2}$ is conditionally convergent.

13. $\lim\limits_{n \to \infty} \frac{n + 1}{7n - 2} = \frac{1}{7}$ so that $\sum\limits_{n=1}^{\infty} (-1)^{n+1} \frac{n + 1}{7n - 2}$ $\underline{\text{diverges}}$ by the nth-term test.

14. $\sum\limits_{n=1}^{\infty} (-1)^{n+1} \dfrac{2^n}{3^n} \approx \dfrac{2}{3} - \dfrac{4}{9} + \dfrac{8}{27} - \dfrac{16}{81} + \dfrac{32}{243} \approx 0.4527$ with an error of magnitude less than $2^6/3^6 < 0.0878$. Since the sign of the first unused term is negative, the sum 0.4527 overestimates the value of the series. In fact, the given geometric series sums to 0.4.

NOTES.

CHAPTER 12 POWER SERIES

12-1 INTRODUCTION.

1. A series of the form

$$\sum_{n=0}^{\infty} a_n x^n = a_0 + a_1 x + a_2 x^2 + a_3 x^3 + \ldots + a_n x^n + \ldots$$

is called a _____.

12-2 TAYLOR POLYNOMIALS.

OBJECTIVE : Find the Taylor Series at $x = a$, or the Maclaurin series, for a given function $y = f(x)$. Assume that $x = a$ is specified and that f has finite derivatives of all orders at $x = a$.

2. If $y = f(x)$ has finite derivatives of all orders at $x = a$, the particular power series

$$f(a) + f'(a)(x - a) + \frac{f''(a)}{2!}(x - a)^2 + \ldots + \frac{f^{(n)}(a)}{n!}(x - a)^n + \ldots$$

is called the _____. If $a = 0$, the series is known as the _____ for f. The Taylor series for a function may or may not converge to the function. This problem is investigated in the next article.

3. If $y = f(x)$ has finite derivatives of order up to and including n, then the polynomial
$$P_n(x) = f(a) + f'(a)(x - a) + \frac{f''(a)}{2!}(x - a)^2 + \ldots + \frac{f^{(n)}(a)}{n!}(x-a)^n$$
is called the nth-degree _____. The graph of this polynomial passes through the point _____, and its first n derivatives match the first n derivatives of _____ at _____. Each nonnegative integer n corresponds to a Taylor polynomial for f at $x = a$, provided the first n derivatives of f exist at $x = a$.

4. Let us find the Taylor polynomials $P_3(x)$ and $P_4(x)$ for the function $f(x) = a^x$, $a > 0$, at $x = 1$. To do this we need to complete the following table:

---

1. formal power series

2. Taylor series for f at $x = a$, Maclaurin series

3. Taylor polynomial of f at $x = a$, $(a, f(a))$, $y = f(x)$, $x = a$

n	$f^{(n)}(x)$	$f^{(n)}(1)$
0	a^x	a
1	$a^x \ln a$	$a \ln a$
2	_____	_____
3	_____	_____
4	_____	_____

Then,

$$P_3(x) = a + a(\ln a)(x - 1) + \frac{a(\ln a)^2}{2!}(x - 1)^2 + \underline{\hspace{3cm}},$$

$$P_4(x) = \underline{\hspace{9cm}}.$$

5.   For the function $f(x) = a^x$ in Problem 4, the Taylor series at $x = 1$ is

$$\sum_{n=0}^{\infty} \underline{\hspace{4cm}}.$$

6.   Let us find the Maclaurin series for the function $f(x) = x^5 + 4x^4 + 3x^3 + 2x + 1$. We need to find the derivatives of f of all orders, and evalute them at $x = 0$:

$f'(x) \quad = \underline{\hspace{3cm}}$, $f'(0) \quad = 2$

$f^{(2)}(x) = \underline{\hspace{3cm}}$, $f^{(2)}(0) = \underline{\hspace{1.5cm}}$

$f^{(3)}(x) = \underline{\hspace{3cm}}$, $f^{(3)}(0) = \underline{\hspace{1.5cm}}$

$f^{(4)}(x) = 120x + 96$, $f^{(4)}(0) = 96$

$f^{(5)}(x) = \underline{\hspace{1.5cm}}$, $f^{(5)}(0) = \underline{\hspace{1.5cm}}$

$f^{(6)}(x) = 0$, $f^{(6)}(0) = 0$

In general, $f^{(k)}(0) = \underline{\hspace{1.5cm}}$ if $k \geq 6$. Thus, the Maclaurin series is

$$\underline{\hspace{9cm}},$$

which simplifies to $1 + 2x + 3x^3 + 4x^4 + x^5$. Therefore, the Maclaurin series for a polynomial expressed in powers of x is the polynomial itself.

---

4.   For $n = k$, $f^{(k)}(1) = a(\ln a)^k$; $\frac{a(\ln a)^3}{3!}(x - 1)^3$,

$a + a(\ln a)(x - 1) + \frac{a(\ln a)^2}{2!}(x - 1)^2 + \frac{a(\ln a)^3}{3!}(x - 1)^3 + \frac{a(\ln a)^4}{4!}(x - 1)^4$

5. $\frac{a(\ln a)^n}{n!}(x - 1)^n$

6. $5x^4 + 16x^3 + 9x^2 + 2$, $\quad 20x^3 + 48x^2 + 18x$, $\quad 0$, $\quad 60x^2 + 96x + 18$, $\quad 18$, $\quad 120$, $\quad 120$, $\quad 0$,

$1 + 2x + 0x^2 + \frac{18}{3!}x^3 + \frac{96}{4!}x^4 + \frac{120}{5!}x^5$

7. Suppose we want to express the polynomial in Problem 6 in powers of (x + 1) instead of powers of x. We find the Taylor series of f at x = _____. From our previous calculations of the derivatives, we find that

$f(-1) = -1$, $f'(-1) = 0$, $f^{(2)}(-1) =$ _____, $f^{(3)}(-1) =$ _____,
$f^{(4)}(-1) =$ _____, $f^{(5)}(-1) =$ _____,   and $f^{(k)}(-1) =$ _____
if $k \geq 6$. Thus the Taylor series of f at x = -1 is

_____,

which simplifies to

$-1 + 5(x + 1)^2 - 3(x + 1)^3 - (x + 1)^4 + (x + 1)^5.$

12-3   TAYLOR'S THEOREM WITH REMAINDER: SINES, COSINES, AND e^x

8. The statement of Taylor's Theorem in the text gives the remainder term as

$R_n(x) =$ _____,

where the number c lies between _____. This remainder term measures the error in the approximation of y = f(x) by the nth-degree Taylor polynomial at _____. Thus, the Taylor series expansion for f(x) will converge to f(x) provided that

_____.

9. This remainder form is very useful because often we can bound the derivative $f^{(n+1)}(c)$ by some constant M: $|f^{(n+1)}(c)| \leq M$. This ensures that $R_n(x)$ converges to _____ as $n \to \infty$.

[OBJECTIVE A]: Using the Maclaurin series for the functions e^x, sin x, and cos x, write the Maclaurin series for functions which are combinations of sines, cosines, exponentials, or powers of x.

10. The Maclaurin series for e^x, sin x, and cos x are

$e^x =$ _____,   sin x = _____,   and cos x = _____.

---

7. -1, 10, -18, -24, 120, 0, $-1 + 0(x + 1) + \frac{10}{2!}(x + 1)^2 - \frac{18}{3!}(x + 1)^3 - \frac{24}{4!}(x + 1)^4 + \frac{120}{5!}(x + 1)^5$

8. $f^{(n+1)}(c)\frac{(x - a)^{n+1}}{(n + 1)!}$, a and x, x = a, $\lim_{n \to \infty} R_n(x) = 0$

9. 0

10. $\sum_{n=0}^{\infty} \frac{x^n}{n!}$, $\sum_{n=0}^{\infty} \frac{(-1)^n x^{2n+1}}{(2n + 1)!}$, $\sum_{n=0}^{\infty} \frac{(-1)^n x^{2n}}{(2n)!}$

11.   Let us find the Maclaurin series for $\sin^3 x$. A trigonometric
       identity gives

$$\sin^3 x = \tfrac{1}{4}(3 \sin x - \sin 3x).$$

We use Maclaurin series for the terms on the right side:

$$3 \sin x = 3x - \frac{3x^3}{3!} + \frac{3x^5}{5!} - \frac{3x^7}{7!} + \cdots ,$$

$$\sin 3x = \underline{\hspace{6cm}} ,$$

$$3 \sin x - \sin 3x = 4x^3 - 2x^5 + \frac{52}{5!} x^7 - \cdots$$

Therefore, $\sin^3 x = \underline{\hspace{5cm}}$

$$= \sum_{n=0}^{\infty} \frac{(-1)^n (3 - 3^{2n+1})}{4(2n+1)!} x^{2n+1}.$$

OBJECTIVE B :   Use the Remainder Estimation Theorem to estimate the
                truncation error when a Taylor polynomial is used to
                approximate a given function. Assume that the function
                has derivatives of all orders.

12.   We will calculate $\cos \sqrt{2}$ with an error less than 10^{-6}. By
       Taylor's Theorem, $\cos \sqrt{2} = \underline{\hspace{6cm}} + R_{2k}(x)$.
       The Remainder Estimation Theorem, with $M = \underline{\hspace{1.5cm}}$, $x = \underline{\hspace{1.5cm}}$,
       and $r = 1$ gives $|R_{2k}| \leq 1 \cdot \underline{\hspace{3cm}}$. By trial we find

       that $\dfrac{(\sqrt{2})^{11}}{11!} = 0.0000011337 > 10^{-6}$ and

$\dfrac{(\sqrt{2})^{13}}{13!} = 0.0000000145 < 10^{-6}$. Thus, we should take $(2k + 1)$

to be at least $\underline{\hspace{1.5cm}}$, or k to be at least 6. With an error
less than 10^{-6}, $\cos \sqrt{2} = 1 - \dfrac{2}{2!} + \dfrac{4}{4!} - \dfrac{8}{6!} + \cdots + \underline{\hspace{2cm}}$

$\uparrow$ last term

$\approx 0.155944.$

13.   Let us determine for what values of $x > 0$ we can replace
       e^x by $1 + x + (x^2/2) + (x^3/3!)$ with an error of magnitude
       less than 5×10^{-5}. In Example 1, page 711 of the text,
       Equation (9d) gives

$$|R_3(x)| < \underline{\hspace{5cm}} .$$

We desire $|R_3(x)| < 5 \times 10^{-5}$. This is the case if
$e^x |x|^4 < \underline{\hspace{2.5cm}}$ or, since $x > 0$, $x + 4 \ln x < -5.116$.
By calculator experimentation this inequality holds if
$0 < x < 0.26$. Thus, for instance,

$$e^{0.1} = 1 + (0.1) + \frac{(0.1)^2}{2} + \frac{(0.1)^3}{6} \approx 1.10517$$

is correct to five decimal places.

---

11. $3x - \dfrac{(3x)^3}{3!} + \dfrac{(3x)^5}{5!} - \dfrac{(3x)^7}{7!} + \cdots ,$ $x^3 - \dfrac{1}{2} x^5 + \dfrac{13}{5!} x^7 - \cdots$

12. $1 - \dfrac{2}{2!} + \dfrac{4}{4!} - \dfrac{8}{6!} + \cdots + (-1)^k \dfrac{2^k}{(2k)!}$, 1, $\sqrt{2}$, $\dfrac{(\sqrt{2})^{2k+1}}{(2k+1)!}$, 13, 6, $\dfrac{64}{12!}$

13. $e^x \cdot \dfrac{x^4}{4!}$, 6×10^{-3}

14. Euler's identity asserts that

$$e^{i\theta} = \underline{\hspace{6cm}}.$$

12-4 EXPANSION POINTS, THE BINOMIAL THEOREM, ARCTANGENTS, AND π.

OBJECTIVE: Use a suitable series to calculate a given quantity to three decimal places. Show that the remainder term does not exceed 5×10^{-4}. (Assume the quantity is the value of a function whose series expansion has been studied in this chapter of the text.)

15. Replacing x by $-x$ in the Taylor series expansion for $\ln(1 + x)$ gives the expansion

$$\ln(1 + x) = \underline{\hspace{6cm}}$$

which is also valid for $|x| < 1$. Subtracting this result from the expansion for $\ln(1 + x)$ gives

$$\ln(1 + x) - \ln(1 + x) = \ln \frac{1 + x}{1 - x} = \underline{\hspace{5cm}}.$$

16. Let N be a positive integer. Then

$$\ln(N + 1) = \ln N + \ln \frac{N + 1}{N}.$$

Now, solve the equation

$$\frac{1 + x}{1 - x} = \frac{N + 1}{N}$$

for x to obtain $x = \underline{\hspace{2cm}}$. Substitution into the result from Problem 15 yields

$$\ln \frac{N + 1}{N} = \underline{\hspace{6cm}}.$$

17. Let's use the result of Problem 15 to calculate $\ln 2$ by setting $N = 1$. Thus,

$$\ln 2 = 2\left(\frac{1}{3} + \frac{1}{3(3)^3} + \frac{1}{3(3)^5} + \frac{1}{7(3)^7} + \frac{1}{9(3)^9} + \frac{1}{11(3)^{11}} + \cdots\right)$$

$$\approx 2(0.3333333 + 0.0123457 + 0.0008230 + 0.0000653 + 0.0000056 + 0.0000005)$$

$$\approx \underline{\hspace{4cm}}.$$

The error satisfies

$$|R_{11}(x)| \le \frac{1}{12} \cdot \frac{|x|^{12}}{1 - |x|} \quad \text{where} \quad x = \frac{1}{2N + 1} = \underline{\hspace{2cm}}.$$

Thus,

$$|R_{11}(x)| \le \frac{1}{12} \cdot \frac{3}{2}|\tfrac{1}{3}|^{12} = 0.00000023252 < 5 \times 10^{-6}.$$

It follows that $\ln 2 \approx 0.69315$, accurate to five decimal places.

---

14. $\cos \theta + i \sin \theta$,   15. $-x - \frac{x^2}{2} - \frac{x^3}{3} - \cdots - \frac{x^n}{n} - \cdots$, $2\left(x + \frac{x^3}{3} + \frac{x^5}{5} + \cdots + \frac{x^{2k-1}}{2k - 1} + \cdots\right)$

16. $\frac{1}{2N + 1}$, $2\left(\frac{1}{2N + 1} + \frac{1}{3(2N + 1)^3} + \frac{1}{5(2N + 1)^5} + \cdots\right)$

17. 0.6931468, $\frac{1}{3}$

18.   Now let's calculate   ln 0.75   using the result in Problem 15.
First use the identity

$$\ln 0.75 = \ln \frac{3}{4} = \ln \frac{3}{2} - \underline{\hspace{3in}}.$$

We can use the calculation for   ln 2   obtained in Problem 17:
ln 2 ≈ 0.69315.   To obtain the first term on the right side of
the previous equation we use the series

$$\ln \frac{N + 1}{N} = 2\left(\underline{\hspace{2.5in}}\right) \quad \text{with} \quad N = 2.$$

Then,
$$\ln \frac{3}{2} = 2\left(\frac{1}{5} + \frac{1}{3(5)^3} + \frac{1}{5(5)^5} + \frac{1}{7(5)^7} + \cdots\right)$$

$$= 2(0.2 + 0.00266667 + 0.000064 + 0.00000183 + \cdots)$$
$$\approx 0.40547.$$

The error satisfies
$$|R_7(x)| \leq \frac{1}{8} \cdot \frac{|x|^8}{1 - |x|}, \quad \text{where} \quad x = \frac{1}{2N + 1} = \underline{\hspace{0.8in}}.$$

Thus,
$$|R_7(x)| \leq \frac{5}{32} \left|\frac{1}{5}\right|^8 < 5 \times 10^{-6}.$$

It follows that
$$\ln \frac{3}{4} \approx 0.40547 - 0.69315 = -0.28768,$$
accurate to five decimal places.

12-5   CONVERGENCE OF POWER SERIES; INTEGRATION, DIFFERENTIATION, MULTIPLICATION, AND DIVISION.

OBJECTIVE A :   Given a power series $\sum\limits_{n=0}^{\infty} a_n x^n$,   find its interval of
convergence.   If the interval is finite, determine
whether the series converges at each endpoint.

19. $\sum\limits_{n=1}^{\infty} \frac{1}{\sqrt{n}\,3^n} x^n$

We apply the ratio test to the series of absolute values, and
find

$$\rho = \lim_{n\to\infty} \left|\frac{x^{n+1}}{\sqrt{n + 1}\,3^{n+1}} \cdot \underline{\hspace{0.7in}}\right| = \lim_{n\to\infty} \frac{\sqrt{n}}{\underline{\hspace{0.7in}}} |x| = \underline{\hspace{0.9in}}.$$

Therefore the original series converges absolutely if
|x| < _____   and diverges if   _____.   When   x = 3,   the
series becomes

$$\sum\limits_{n=1}^{\infty} \underline{\hspace{0.9in}}, \quad \text{the p-series with} \quad p = \underline{\hspace{0.9in}};$$

this series _____.   When   x = -3,   the series becomes

---

18. ln 2, $\dfrac{1}{2N + 1} + \dfrac{1}{3(2N + 1)^3} + \dfrac{1}{5(2N + 1)^5} + \cdots$, $\dfrac{1}{5}$

$$\sum_{n=1}^{\infty} \underline{\hspace{3cm}},$$

and this series _____, by Leibniz's Theorem. Therefore, the interval of convergence of the original power series is _____.

20. $\displaystyle\sum_{n=1}^{\infty} \frac{2^n}{n(3^{n+2})} x^{n+1}$

The power series converges for $x = 0$. For $x \neq 0$, we apply the root test to the series of absolute values, and find

$$\rho = \lim_{n \to \infty} \sqrt[n]{\frac{2^n |x|^n |x|}{n \cdot 3^n \cdot 3^2}} = \lim_{n \to \infty} \underline{\hspace{4cm}}$$

$$= \frac{2|x| \cdot 1}{\underline{\hspace{1.5cm}}} < 1, \quad \text{if} \quad |x| < \underline{\hspace{1.5cm}}.$$

Therefore, the original series converges absolutely if $|x| < 3/2$ and diverges if $|x| > 3/2$. When $x = 3/2$, the series becomes

$$\sum_{n=1}^{\infty} \frac{2^n}{n\left(3^{n+2}\right)} \left(\frac{3}{2}\right)^{n+1} = \sum_{n=1}^{\infty} \underline{\hspace{2cm}},$$

and this series _____. When $x = -3/2$, the series becomes

$$\sum_{n=1}^{\infty} \frac{(-1)^{n+1}}{6n},$$

and this series _____, by Leibniz's Theorem. Therefore, the interval of convergence of the original power series is _____.

21. $\displaystyle\sum_{n=1}^{\infty} [\sin (5n)](x - \pi)^n$

For every value of x, $|[\sin (5n)](x - \pi)^n| \leq |x - \pi|^n$. The geometric series $\displaystyle\sum_{n=1}^{\infty} |x - \pi|^n$ converges if _____ and diverges if _____. Therefore, by the comparison test, the original series converges absolutely if _____.
Suppose $|x - \pi| = 1$. Then the series becomes,

$$\sum_{n=1}^{\infty} \sin (5n) \quad \text{or} \quad \sum_{n=1}^{\infty} \underline{\hspace{3cm}}.$$

However, $\displaystyle\lim_{n \to \infty} \sin (5n)$ fails to exist, so neither of these series can converge. We conclude that the interval of convergence of the original series is _____.

---

19. $\dfrac{\sqrt{n}\ 3^n}{x^n}$, $3\sqrt{n+1}$, $\frac{1}{3}|x|$, 3, $|x| > 3$, $\frac{1}{\sqrt{n}}$, $\frac{1}{2}$, diverges, $\dfrac{(-1)^n}{\sqrt{n}}$, converges, $-3 \leq x < 3$

20. $\dfrac{2|x| \cdot \sqrt[n]{|x|}}{n\sqrt{n} \cdot 3 \cdot \sqrt[n]{9}}$, $1 \cdot 3 \cdot 1$, $\frac{3}{2}$, $\frac{1}{6n}$, diverges, converges, $-\frac{3}{2} \leq x < \frac{3}{2}$

21. $|x - \pi| < 1$, $|x - \pi| \geq 1$, $|x - \pi| < 1$, $(-1)^n \sin (5n)$, $\pi - 1 < x < \pi + 1$

OBJECTIVE B : Given a power series $f(x) = \sum a_n x^n$, find the power series for $f'(x)$.

22.   In Example 7, on page 733 of the text, it is given that
$$\frac{1}{1 + t^2} = 1 - t^2 + t^4 - t^6 + \cdots, \quad \text{for} \quad -1 < t < 1.$$

Therefore, using the term-by-term differentiation theorem,

$$\frac{-2t}{\left(1 + t^2\right)^2} = \underline{\hspace{4cm}}, \quad \text{for} \quad \underline{\hspace{3cm}}.$$

OBJECTIVE C : If f is a function having a known power series $f(x) = \sum a_n x^n$, use the series and a calculator to estimate the integral $\int_0^b f(x)\,dx$, assuming that b lies within the interval of convergence.

23.   Let us find $\int_0^{0.2} \cos \sqrt{x}\, dx$ accurate to five decimal places. Now,
$$\cos x = 1 - \frac{x^2}{2!} + \frac{x^4}{4!} - \frac{x^6}{6!} + \frac{x^8}{8!} - \cdots,$$
so the power series for $\cos \sqrt{x}$ is given by

$$\cos \sqrt{x} = \underline{\hspace{6cm}}, \quad x \geq 0.$$
Thus, using term-by-term integration,

$$\int_0^{0.2} \cos \sqrt{x}\, dx = \underline{\hspace{6cm}} \Big]_0^{0.2}$$
$$= 0.2 - \frac{0.04}{4} + \frac{0.008}{72} - \frac{0.0016}{2880} + \frac{0.00032}{201600} - \cdots$$
$$\approx 0.2 - 0.01 + 0.00011 - 0.00000056 + \cdots$$
Hence, $\int_0^{0.2} \cos \sqrt{x}\, dx \approx \underline{\hspace{2.5cm}}$

with an error of less than 5×10^{-6}.

12-6   INDETERMINATE FORMS.

OBJECTIVE : Use series to evaluate the limit $\lim_{x \to a} \dfrac{f(x)}{g(x)}$, at a point a where $f(x)$ and $g(x)$ are both zero. Assume that the functions f and g have series expansions in powers of $x - a$ that converge in some interval $|x - a| < \delta$.

---

22. $-2t + 4t^3 - 6t^5 + \cdots, \ -1 < t < 1$

23. $1 - \dfrac{x}{2!} + \dfrac{x^2}{4!} - \dfrac{x^3}{6!} + \dfrac{x^4}{8!} - \cdots, \ x - \dfrac{x^2}{2 \cdot 2!} + \dfrac{x^3}{3 \cdot 4!} - \dfrac{x^4}{4 \cdot 6!} + \dfrac{x^5}{5 \cdot 8!} - \cdots, \ 0.19011$

24. $\lim\limits_{x \to 0} \dfrac{e^{2x} - 1}{x}$

The Maclaurin series for e^{2x}, to terms in x^3, is

$$e^{2x} = \sum_{n=0}^{\infty} \frac{(2x)^n}{n!} = \underline{\hspace{5cm}}.$$

Hence, $e^{2x} - 1 = 2x\left(\underline{\hspace{4cm}}\right),$ and

$$\lim_{x \to 0} \frac{e^{2x} - 1}{x} = \lim_{x \to 0} 2\left(1 + x + \frac{2}{3}x^2 + \dots\right) = \underline{\hspace{2cm}}.$$

25. $\lim\limits_{x \to 0} \dfrac{\tan x - x}{x - \sin x}$

The Maclaurin series for $\tan x$ and $\sin x$, to terms in x^5, are

$$\tan x = x + \frac{x^3}{3} + \frac{2x^5}{15} + \dots, \quad \sin x = \underline{\hspace{4cm}}.$$
Hence,

$$\tan x - x = \frac{x^3}{3}\left(1 + \frac{2}{5}x^2 + \dots\right) \quad \text{and}$$

$$x - \sin x = \frac{x^3}{3}\left(\underline{\hspace{3cm}}\right).$$

Therefore,

$$\lim_{x \to 0} \frac{\tan x - x}{x - \sin x} = \lim_{x \to 0} \frac{\left(\underline{\hspace{3cm}}\right)}{\left(\frac{1}{2} - \frac{1}{40}x^2 + \dots\right)} = \underline{\hspace{2cm}}.$$

12-7   A COMPUTER MYSTERY.

26. Why did the computer or calculator produce the answer 1 for calculating $\left(1 + \frac{1}{n}\right)^n$ when $n = 10^{13}$, $n = 10^{14}$, etc., on page 743?

---

24. $1 + 2x + \dfrac{4x^2}{2!} + \dfrac{8x^3}{3!} + \dots$, $1 + x + \dfrac{2}{3}x^2 + \dots$, 2

25. $x - \dfrac{x^3}{3!} + \dfrac{x^5}{5!} - \dots$, $\dfrac{1}{2} - \dfrac{1}{40}x^2 + \dots$, $1 + \dfrac{2}{5}x^2 + \dots$, 2

26. The number $\frac{1}{n}$ is so small that when added to 1 the computer cannot show any digits beyond the 12th decimal place. Thus the computer effectively sets $1 + \frac{1}{n}$ equal to 1, and $1^n = 1$.

CHAPTER 12 SELF-TEST

1. Find the Taylor series of $f(x) = \sqrt{x}$ at $a = 9$. Do not be concerned with whether the series converges to the given function f.

2. Find the Maclaurin series for the function $f(x) = x \ln (1 + x^2)$ using series that have already been obtained in the Thomas/Finney text.

3. Use series to estimate the number $e^{-1/3}$ with an error of magnitude less than 0.001.

4. Use series to evaluate the following limits.

    (a) $\displaystyle\lim_{x \to 0} \frac{\sin x - x \cos x}{x^3}$ (b) $\displaystyle\lim_{x \to 0} \frac{\ln(1 - 2x)}{\tan \pi x}$

5. Find the first three nonzero terms in the Maclaurin series for the function $f(x) = \sec^2 x$ using the Maclaurin series for tan x.

6. (Calculator) Use series and a calculator to estimate the integral

    $$\int_0^{0.5} \cos x^2 \, dx$$

    with an error of magnitude less than 0.0001.

SOLUTIONS TO CHAPTER 12 SELF-TEST

1. We calculate the derivatives of $f(x) = \sqrt{x}$, and evaluate f and these derivatives at $a = 9$:

$$
\begin{aligned}
f(x) &= \sqrt{x} & f(9) &= 3 \\
f'(x) &= \tfrac{1}{2} x^{-1/2} & f'(9) &= \tfrac{1}{6} \\
f^{(2)}(x) &= (-1)\left(\tfrac{1}{2}\right)\left(\tfrac{1}{2}\right) x^{-3/2} & f^{(2)}(9) &= -\tfrac{1}{108} \\
f^{(3)}(x) &= (-1)^2\left(\tfrac{1}{2}\right)\left(\tfrac{1}{2}\right)\left(\tfrac{3}{2}\right) x^{-5/2} & f^{(3)}(9) &= \tfrac{1}{648} \\
f^{(4)}(x) &= (-1)^3 \frac{3 \cdot 5}{2^4} x^{-7/2} & f^{(4)}(9) &= -\tfrac{5}{11664} \\
&\quad\vdots & &\quad\vdots \\
f^{(k)}(x) &= (-1)^{k+1} \frac{3 \cdot 5 \cdots (2k-3)}{2^k} x^{-(2k-1)/2} \\
& & f^{(k)}(9) &= (-1)^{k+1} \frac{3 \cdot 5 \cdots (2k-3)}{2^k \, 3^{2k-1}}
\end{aligned}
$$

Therefore, the Taylor series for $f(x) = \sqrt{x}$ at $a = 9$ is

$$3 + \tfrac{1}{6}(x-9) - \tfrac{1}{216}(x-9)^2 + \cdots + (-1)^{k+1} \frac{3 \cdot 5 \cdots (2k-3)}{2^k \, 3^{2k-1} \cdot k!} (x-9)^k + \cdots$$

2. $\ln (1 + x) = x - \dfrac{x^2}{2} + \dfrac{x^3}{3} - \dfrac{x^4}{4} + \ldots \, ,$ $-1 < x \le 1$

   $\ln (1 + x^2) = x^2 - \dfrac{x^4}{2} + \dfrac{x^6}{3} - \dfrac{x^8}{4} + \ldots \, ,$ $-1 < x \le 1$

   $x \ln (1 + x^2) = x^3 - \dfrac{x^5}{2} + \dfrac{x^7}{3} - \dfrac{x^9}{4} + \ldots \, ,$ $-1 < x \le 1$

   or, in closed form, $x \ln (1 + x^2) = \displaystyle\sum_{n=0}^{\infty} (-1)^n \dfrac{1}{n + 1} x^{2n+3}$,   valid for

   all x satisfying $-1 < x \le 1$.

3. $e^{-1/3} = 1 - \dfrac{1}{3} + \dfrac{(-1/3)^2}{2!} + \dfrac{(-1/3)^3}{3!} + \dfrac{(-1/3)^4}{4!} + \ldots$

   By trial, $\dfrac{(-1/3)^4}{4!} < 0.00052$ and $\dfrac{(1/3)^3}{3!} > 0.001$.

   Since the series is an alternating series,

   $e^{-1/3} \approx 1 - \dfrac{1}{3} + \dfrac{1/9}{2!} - \dfrac{1/27}{3!} = 0.71605$ with an error in magnitude

   less than 0.00052.

4. (a) The Maclaurin series for $\sin x$ and $x \cos x$, to terms in
         x^7, are

         $\sin x = x - \dfrac{x^3}{3!} + \dfrac{x^5}{5!} - \dfrac{x^7}{7!} + \ldots \, ,$

         $x \cos x = x - \dfrac{x^3}{2!} + \dfrac{x^5}{4!} - \dfrac{x^7}{6!} + \ldots \, .$ Hence,

   $\sin x - x \cos x = x^3\left(\dfrac{1}{2!} - \dfrac{1}{3!}\right) + x^5\left(\dfrac{1}{5!} - \dfrac{1}{4!}\right) + x^7\left(\dfrac{1}{6!} - \dfrac{1}{7!}\right) + \ldots \, ,$

         and

   $\displaystyle\lim_{x \to 0} \dfrac{\sin x - x \cos x}{x^3} = \lim_{x \to 0} \left[\left(\dfrac{1}{2!} - \dfrac{1}{3!}\right) + x^2\left(\dfrac{1}{5!} - \dfrac{1}{4!}\right) + \ldots\right]$

   $= \dfrac{1}{2} - \dfrac{1}{6} = \dfrac{1}{3}.$

   (b) The Maclaurin series for $\ln (1 - 2x)$ and $\tan \pi x$ are

        $\ln (1 - 2x) = -2x - \dfrac{(2x)^2}{2} - \dfrac{(2x)^3}{3} - \ldots \, ,$

        $\tan \pi x = \pi x + \dfrac{(\pi x)^3}{3} + \dfrac{2(\pi x)^5}{15} + \ldots$

        Hence,

   $\dfrac{\ln (1 - 2x)}{\tan \pi x} = \dfrac{-x\left(2 + 2x + \dfrac{8x^2}{3} + \ldots\right)}{x\left(\pi + \dfrac{\pi^3 x^2}{3} + \dfrac{2\pi^5 x^4}{15} + \ldots\right)}$

        and

   $\displaystyle\lim_{x \to 0} \dfrac{\ln(1 - 2x)}{\tan \pi x} = \lim_{x \to 0} \dfrac{-(2 + 2x + \ldots)}{\left(\pi + \dfrac{\pi^3 x^2}{3} + \ldots\right)} = -\dfrac{2}{\pi}.$

5.   The Maclaurin series for tan x, through the first three nonzero terms, is

$$\tan x = x + \frac{x^3}{3} + \frac{2x^5}{15} + \dots$$

Hence, $\sec^2 x = \dfrac{d}{dx} \tan x = 1 + x^2 + \dfrac{2}{3}x^4 + \dots$.

6.   The Maclaurin series for $\cos x^2$ is

$$\cos x^2 = 1 - \frac{x^4}{2!} + \frac{x^8}{4!} - \frac{x^{12}}{6!} + \dots + (-1)^k \frac{x^{4k}}{(2k)!} + \dots$$

Hence,

$$\int_0^{0.5} \cos x^2 \, dx = x - \frac{x^5}{5 \cdot 2!} + \frac{x^9}{9 \cdot 4!} - \frac{x^{13}}{13 \cdot 6!} + \dots \Big]_0^{0.5}$$

$$\approx 0.5 - 0.00313 + 0.0000090 - \dots = 0.49687,$$

with an error in magnitude less than 0.000009 because the series is an alternating series.